PARASITE LIVES

John Frederick Adrian Sprent

PARASITE LIVES

Papers on Parasites, Their Hosts and Their Associations

To Honour J. F. A. Sprent

Edited by
Mary Cremin, Colin Dobson & Douglas E. Moorhouse

University of Queensland Press
St Lucia • London • New York

First published 1986 by University of Queensland Press
Box 42, St Lucia, Queensland, Australia

Typeset by Press Etching Pty Ltd, Brisbane
Printed in Australia by Dominion Press—Hedges & Bell

Distributed in the UK and Europe by University of Queensland Press
Dunhams Lane, Letchworth, Herts. SG6 1LF England

Distributed in the USA and Canada by University of Queensland Press
250 Commercial Street, Manchester, NH 03101 USA

Cataloguing in Publication Data

National Library of Australia

Parasite lives.

Bibliography.

1. Sprent, J. F. A. (John Frederick Adrian), 1915- . 2. Parasites — Addresses, essays, lectures. 3. Parasitology — Addresses, essays, lectures. 4. Host-parasite relationships — Addresses, essays, lectures. I. Cremin, Mary, 1925– . I. Dobson, Colin, 1937– . III. Moorhouse, Douglas E. (Douglas Edward), 1925- . IV. Sprent, J. F. A. (John Frederick Adrian), 1915- .

591.52'49

British Library (data available)

Library of Congress

Parasite lives.

Bibliography: p.

1. Parasites. 2. Host-parasite relationships. 3. Parasitism. I. Cremin, Mary, 1925- . II. Dobson, Colin, 1937- . III. Moorhouse, Douglas, E. (Douglas Edward), 1925- .

QL757.P275 1986 591.5'249 86-11334

ISBN 0 7022 2041 8
ISBN 0 7022 2042 6 (pbk.)

Contents

Part Four: Host-Parasite Interactions

Preface

Our book honours Emeritus Professor John Frederick Adrian Sprent C.B.E., F.A.A. in his seventy-first year and the occasion of the Sixth International Congress of Parasitology which he worked hard to bring to the University of Queensland, Brisbane, under the auspices of the Australian Society for Parasitology.

John Sprent has done much that identifies the University of Queensland and Brisbane as important to the development of parasitology through his skills and long experience as an educator, administrator and research worker. Equally he is well known for the warmth of his personality, innate humanity and his good fellowship. It is appropriate, therefore, that his students and colleagues should wish to express their debt to him as a mentor by contributing the fruits of their own research to our festschrift. Many of his friends and associates were unable, for various reasons, to contribute but fête him with their warm felicitations.

I believe that the title of our book *Parasite Lives* illuminates all John's parasitological curiosity. It encapsulates his critical approach to classical taxonomy that is so well tempered by his elegant studies on the biology and life cycles of parasites; his thoughts on the interactions of parasites with their hosts that have such strong appeal to younger "immunoparasitologists"; his concern to construct meaningful phylogenies that are illustrated by considered zoogeographies of parasites and hosts and opinions on the influence on mankind of host-parasite relationships; and finally his unflagging allegiance to Darwinian evolution that edifies and entertains us. It follows naturally that a man with such catholic views in parasitology has produced students and followers, and gathered colleagues and visitors about him, that specialize in many of the varied aspects of parasitology. Our book is structured to show this association and attempts to demonstrate the gamut of his interests.

The editors append a biography of John Sprent which gives the dates of the distinctions he has received and a list of his publications. We were fortunate also to obtain the text of an address, "Parasitological Reminiscences", which he gave to local members of the Australian Society for Parasitology, 18 May 1982, which contains personal insights to his career and interests. The talk was invited by three postgraduate students, Shirley Butler, Tom Cribb and Malcolm Jones, to celebrate the centenary of the publication of the life cycle of *Fasciola hepatica*; the text makes an interesting opening to the festschrift.

John, all your friends wish you well for the future. May you long continue as editor-in-chief of the *International Journal for Parasitology* and may your research flourish here in the Department of Parasitology.

Colin Dobson

Acknowledgments

The editors gratefully acknowledge the assistance of Drs K. C. Bremner, R. M. Cable, L. R. G. Cannon, H. M. D. Hoyte, R. M. Overstreet, and J. C. Pearson, and Mr L. A. Y. Johnston who carefully reviewed individual papers. Betty M. Siddell, Elizabeth A. Weston, Sally Roth, Margaret E. Owen, and Graham Leatch gave valuable editorial and secretarial assistance. We thank the staff of the University of Queensland Press for their cheerful and professional service that smoothed the progress of putting the book together particularly in the last months before the Sixth International Congress of Parasitology. We are pleased to acknowledge the generous financial support given to us by the University of Queensland.

Manuscripts were received between 29 September 1982 and 14 January 1983. Revisions were received between 2 April 1984 and 29 August 1985.

Part One

Parasitological Reminiscences

An Address by J. F. A. Sprent Delivered to Members of the Australian Society for Parasitology on 18 May 1982

Early in the year, there was a knock on my door and there stood three postgraduate students. Bearing in mind that it is 1982 — the centenary of the publication on the development in *Limnaea truncatula* of *Fasciola hepatica* — they asked whether I would give a talk on some aspect of parasitology of my choice. Of course I felt very honoured to be asked and set about searching for a topic. I soon found that 1882 and thereabouts were important years for many reasons.

It is appropriate that I begin with the liver fluke. The winter of 1879–80 in the British Isles had been very wet and about 10% of all sheep had died of liver rot. It was known that the disease was caused by *Fasciola hepatica*; it was known also that in other trematodes there is a snail stage of development producing infective cercariae. It was suspected that a snail stage was involved with *F. hepatica*, but the snail was not known. In April 1880 there was a letter in *The Times* from T.S. Cobbold to the effect that Professor Leuckart considered *L. truncatula* to be the culprit. A month or two later, the Royal Agricultural Society of England offered a grant to Dr George Rolleston to investigate the disease. He was unable to take up the grant and it was offered to A.P. Thomas, Demonstrator in the Oxford University Museum. In April 1881 he published a description of a cercaria in *L. truncatula*, but he had not proved that it was *F. hepatica* by infecting snails from the eggs. However, after much difficulty over finding more snails for this purpose, he wrote up his results in an article in *Nature* (19 October 1882) in which he described the form and behaviour of the miracidium, reported how he had infected *L. truncatula*, demonstrated the host-specificity of the mollusc phase, and described the development of the sporocyst, redia, and cercaria. This work was published in greater detail in the *Quarterly Journal of Microscopical Science* in 1882. By this time several papers on the subject had appeared by K.G.F.R. Leuckart; in December 1881 he reported that he regarded *L. pereger* as the host; in early 1882 he reported that he had found three cercariae in *L. trunculata*, but he evidently did not yet know which of them was *F. hepatica*. In October 1882 he published a fuller account and differentiated the cercariae. Neither Thomas nor Leuckart succeeded in infecting sheep with *F. hepatica*; this was demonstrated by Adolph Lutz in Hawaii in 1892.

Little more was heard of A. P. Thomas in the parasitological world. He went to New Zealand and became Sir Algernon Thomas. In contrast, Leuckart became, in a sense, the founder of modern parasitology. His students included such men as C.W. Stiles and H.B. Ward; among the latter's students nine became presidents of the American Society of Parasitologists.

As I was browsing through the volume of *Nature* for 1882 I came across a description of Darwin's funeral in the May issue. The rancour of the sixties over natural selection was all forgotten, his remains were received by the Archbishop of Canterbury and laid to rest in Westminster Abbey near Isaac Newton. I was reminded of the hymn — All things bright and beautiful, All creatures great and small, All things wise and wonderful, The Lord God made them all — I wondered whether they sang it at the funeral just to show that, as good church people, they did not believe all that nonsense Darwin had written about natural selection.

Fortunately, the less desirable forms of animal life were not regarded by the Church as befitting objects for "God's Almighty Hand", so that parasitologists have been less involved than other biologists with the "Creation versus Evolution" controversy (but you should read the *Newsletter of the American Society of Parasitologists*, vol. 4, no. 1, March 1982).

There were three main theories on the origins of worms in the human body: (1) they were derived from free living animals which had been swallowed; (2) they were spontaneously generated; and (3) they were preformed in the embryo. However they got there, the Church tended to regard parasites as the work of the Devil rather than God. If they were spontaneously generated, it was from over-eating or some other form of fleshly indulgence. If they were preformed, blame was placed upon the unfortunate Eve who, for her "original sin", was destined to pass to every human infant an inbuilt prepackaged bundle of embryonic worms. Cobbold (1879) summed this up better than I can: "A superstitious age sought to interpret their presence as having some connection with human wrong-doing . . . Possibly it is only by accepting the hypothesis of natural selection that we can escape the somewhat undignified conclusion that the entozoa were expressly created to dwell in us and that we were designed and destined to entertain them."

Despite the detection by Tyson of reproductive organs and eggs in the large roundworm of man, despite the observations of Pallas on developing eggs, despite the discovery by various observers of miracidia and coracidia in eggs of platyhelminths, the doctrine of spontaneous generation lasted well into the 19th century. It was Küchenmeister in the 1850s, by demonstrating that tapeworms grew from cysticerci, who put the lid on its coffin and Pasteur drove in the last few nails.

Spontaneous generation was dead: Long live parasitology! Parasitology was born when it was realized that people get "infected" with parasites rather than spontaneously generating them. Indeed, parasitology was born in 1882 when the term "Parasitology" first appeared in a dictionary. The term "parasitism" (in the biological sense) had been used shortly before, when the host-parasite relationship was compared with commensalism and mutualism by Van Beneden in his book *Animal Parasites and Messmates*.

These were stirring times also in other respects. It was in 1882 that Metchnikov published the first of his series of papers on phagocytosis, and Pasteur showed that the anthrax bacillus, after growth in chicken broth at 40°C lost its virulence and could be used for immunization. The most important parasitological discoveries of those times were those which associated parasites with the cause of hitherto unexplained disease and showed how they were transmitted. Such discoveries were made by different people, in different places, at different times. For example, Lavaran in 1882 reported from Algiers that he had found small clear spots in the erythrocytes of patients with malaria; he described how the spots grew, acquired pigment, and gradually filled the cell and burst it; he assumed that they were derived from the damp air of marshes. In New York, in 1882, at the International Sanitary Conference, Carlos Finlay, on leave from Havana, was enunciating his theory on the possible transmission of yellow fever by mosquitoes, and a young man, Ronald Ross by name, having just received his diploma from the Society of Apothecaries, was on his way to serve in the Madras Medical Service.

In the next few decades parasitology flourished enormously, through such men as Ross, Manson, Bruce, Looss, and others; all were dedicated men, motivated towards helping the sufferers of parasitic disease in the tropics. Their support came mostly through industrial enterprise. "*Tutamen zonae torridae*" is the motto of the Royal Society of Tropical Medicine and Hygiene, but one wonders whether the hurry was not so much to cure and aid the sick, but rather to enable them to work harder, planting tea and coffee, extracting palm oil, mining, and excavating the Panama Canal.

As it turned out, the publication of this period which was most interesting to me personally was a book. It was *Island Life* by Alfred Russell Wallace, a popular supplement to his *Geographical Distribution of Animals*, published in 1876. This book had perchance found its way into my grandfather's bookshelves. It was a stranger there, amid a miscellany of biographies of eminent clergymen. I will refer later to this wonderful book.

My father wished me to have a classical education and at considerable sacrifice sent me to Shrewsbury School in Shropshire, where it was intended that I should take Ancient Languages in preparation for "Greats" at Oxford. I regret to say that I was a disappointment. None of my teachers, not even the shade of Charles Darwin, a former pupil, could light in me a spark of enthusiasm for learning. I never achieved a higher position in my classes than bottom or one from bottom. My father died when I was sixteen and I had to leave school and find a job. It was at the time of the Great Depression, but I was lucky to be given a job as an office boy in a firm of cigarette manufacturers in London. But here again my performance was disappointing. After two years I was still the most junior member of the firm, and I was given to understand that my services were likely to be dispensed with.

I tried for a short service commission as a pilot in the Royal Air Force but failed the medical. Then one evening I was strolling with a friend from the office across Hungerford Bridge to Waterloo Station to catch his train to Surbiton. He told me how he had wanted to go to university, but it was now too late. He asked whether I would be willing to do what he had wanted to do and he offered to pay my fees. At first I thought he was joking, but he was serious. He said that he knew that I could do it and wanted to help me to do so.

To cut a long story short, I accepted his offer and through his confidence, a new world was opened to me. With his encouragement and help, I received my diploma from the Royal College of Veterinary Surgeons five years later. This would not have been possible without the support of my mother and family and especially that of my wife, Muriel, whom I married in 1937. It was the zoological aspects which interested me most, and two books in particular had influenced me in deciding to seek work in the tropics. One was the above-mentioned *Island Life*, the other was *Arrowsmith* by Sinclair Lewis. I applied for and was awarded a scholarship from the Colonial Office. The agreement was that I should take a degree in Zoology and afterwards join the Colonial Service for work in tropical Africa.

Meanwhile, the war had started and when in 1942 I left England for the first time, the battle of the Atlantic was in full swing. It took six weeks to reach the coast of West Africa and several more weeks to reach the tiny veterinary station at Vom in northern Nigeria. On arrival I was greatly disappointed to learn that all veterinary officers were expected to devote their first term of service to Rinderpest immunization. Parasites however came to my rescue. I had a few days at Vom and I started examining water from a ditch near my house and to my surprise found it to be seething with forked-tail cercariae. I told the director of the station about the possibility of schistosomiasis and he was quite alarmed. He told me to stay at Vom a bit longer to write a report on it. A few days later one of the stock farm cattle died and at autopsy I found it was heavily infected with hookworms and very anaemic. Again I was asked to write a report and so the weeks turned into months and I spent two years working mainly on hookworms in cattle. When I returned to England on leave, I submitted my report as an external Ph.D. thesis to the University of London.

While on leave I had the good fortune to seek the advice of three people who influenced me greatly by their work and ideas, Professor R.T. Leiper, Dr H.A. Baylis, and Dr E.L. Taylor. The war had ended and with their help I was given the opportunity to spend two years at the University of Chicago with Dr W.H. Taliaferro. I remember how exhilarating and refreshing it was to arrive in the United States. The famous people strolling about the campus, Fermi, Urey, and Oppenheimer, were working right opposite the laboratory under the football stadium where the first atomic pile had been constructed. I remember the first meeting I attended of the American Society of Parasitologists, conversing for the first time with people like Cort, Ackert, Schwartz, Stunkard, and Chitwood, and hearing Stoll's Presidential Address "This Wormy World".

At Chicago I became particularly interested in the work which José Oliver-Gonzalez had done there with the component tissues of *Ascaris suum*. He had found that they contained agglutinogenic factors. I was interested in finding which tissues contained antigens functional in promoting resistance to infection. When mice are given a first infection with eggs of *A. suum*, the larvae move from the liver to the lungs, are mostly in the lungs at 7 days, and disappear up the trachea by 8 days. On a second infection many are found to be retained in the liver by 7 days. I used this "liver-lung ratio" as a measure of resistance, using the component parts of the body of *A. suum* as antigens. But it was only the metabolic products in which I could detect any protective function.

In 1947, I was offered a position in the Ontario Research Foundation in Toronto, Canada, to work on diseases of fur-bearing animals. I had occasion to autopsy a wide variety of animals which I had never encountered before: bobcat, lynx, wolf, marten, bear, raccoon, skunk, beaver, otter, and many others. I found many "*Ascaris*" species which were new to me. I decided to continue my immunity studies and to use these indigenous species instead of *A. suum*. To my surprise I found that the larvae of each species had a characteristic migratory pattern in rodents, each quite different from that of *A. suum*. The larvae migrated to muscles, eye, brain, heart, and many other tissues and could remain there in a quiescent state for years. I realized that here in the Holarctic forests ascaridoid nematodes dwelt in two tiers of hosts, the larvae in the prey (mainly rodents) and the adults in the predators (Carnivora). Thus, their life history patterns resemble those known for ascaridoids of seals and other aquatic predators.

When I came to Australia I was able to observe that similar life history patterns prevailed among ascaridoids of reptiles. Indeed it became evident that the life history patterns of all but a few ascaridoids of herbivorous hosts are enacted within a food pyramid, in many instances with an invertebrate first intermediate host at the base of the pyramid, an invertebrate-eating second intermediate host in the middle, and a dominant predator at the peak. Host-specificity is wide at the base and narrow at the peak. By this time, life history patterns and migratory behaviour of ascaridoids had directed my interests away from immunology. At Chicago, emphasis was mainly on immunological aspects related to diagnosis and immunization. The macrophage was the centre of interest and especially the influence of antigen on antibody production. The names which figured predominantly in my reading were Bordet, Landsteiner, Haurowitz, and Pauling. The names of Lederberg, Talmage, Jerne, Medawar, and Burnet were yet to come.

With the publication of *The Clonal Selection Theory of Acquired Immunity* in 1959 my interest in immunology was rekindled, because clonal selection seemed to provide much more scope than the template concepts in relation to the evolution of parasitism. This theme was the gist of an essay on the implications of Burnet's theory which I wrote for the Darwin Centenary Symposium in Melbourne, 1959. I asked the question as to whether host-parasite relationships in which there was minimal immune response had been selected over evolutionary time, i.e., hosts were selected which were immunologically non-competent to particular parasite antigens, and parasites had been selected which had tailored their excretory and secretory products to conform with host molecules. I envisaged a form of immunosuppression acquired over evolutionary time which I termed "adaptation tolerance".

By comparing host reaction and growth of ascaridoid larvae in indigenous and non-indigenous Australian rats, it appeared that the older the relationship, the less reactive is the the host, and the more actively the larvae grow.

But yet again my interest was diverted from immunology, because in 1960 two papers came to my notice. One was a reclassification of ascaridoids based on the excretory system by Hartwich. The other was an essay on the evolution of Ascaridoidea by Osche.

Osche had been investigating the development of an ascaridoid in the European blackbird. He had collected various developmental stages and noted the changes in lip structure as development proceeded. He considered that there was a similarity of lip structure between the developmental stages and certain genera which Hartwich had depicted. Accordingly he arranged the genera according to this lip structure in a "phylogenetic series" and postulated that the series was recapitulated in development. Those with the smallest lips in the adult were, according to Osche, the most primitive. By associating the hosts within his phylogenetic series, Osche postulated that ascaridoids originated in sharks and that they had subsequently radiated among the vertebrates by parallel evolution, a process taking some 200 million years.

This idea of an aquatic origin in sharks was difficult to reconcile with Chitwood's view that the origin of ascaridoids was in terrestrial hosts, probably from forms in insects. Accordingly in 1962 I suggested that it might be interesting to investigate whether ascaridoids might have originated from primarily marine aphasmidian nematodes. This suggestion was received with singular lack of enthusiasm by nematologists, such as Chabaud, Anderson and Inglis, so that there seemed to be no alternative but to re-examine Hartwich's classification and Osche's phylogenetic system.

Without fossils, the parasitologist is at a disadvantage when it comes to tracing evolutionary pathways. On the other hand, going back again to 1882, the evidence of the geographical distribution of animals accumulated by Wallace in *Island Life* turned out to be as significant in supporting the Darwin-Wallace hypothesis as was the fossil evidence provided by Lyell.

It occurred to me that one way to test Osche's hypothesis about the age of ascaridoid-vertebrate relationships was to do with this parasite group what Wallace did with the free-living animals, i.e. determine the host and geographical distribution of all ascaridoid species so that this data could be related to the overall distribution of their hosts and to the appearance and disappearance of geographical barriers. It was, however, necessary to heed Wallace's warning that one must be certain of true affinity of species before drawing conclusions about their distributions.

As the work is by no means completed, it is possible to give only tentative conclusions.

With regard to host distribution, the following broad generalizations can be listed:

(1) The host distribution of ascaridoids among the vertebrates is not indicative of an ancient parallel ramification among the vertebrate classes. On the contrary, among each class of the vertebrates the present day host distribution of ascaridoids, except among the crocodilians, manifests a characteristic sporadic pattern evidently determined mainly by feeding habits.

(2) The most primitive members of the host classes are either not infected or are infected by a disparate assortment of unrelated genera, indicating secondary (souvenir) rather than primary (heirloom) exploitation.

(3) Although a few herbivorous hosts are infected, e.g., *Testudo, Castor, Ailuropoda, Equus, Bos* there is no indication of radiation among related herbivorous host species, suggesting secondary adaption.

(4) Their host distribution strongly suggests that ascaridoids have become incorporated into particular food chains, they tend to occur in both larval and adult forms in lowly predators (especially cyprinoid fishes, frogs and rats) and as adults in dominant predators (sharks, snakes, crocodiles, hawks, cormorants, toothed whales and most of the Carnivora).

In considering the geographical distribution of ascaridoids as a clue to their evolutionary history one is interested particularly in those whose hosts are likely to have been restricted by geographical barriers. For this reason, ascaridoids of freshwater fishes, amphibians, terrestrial and freshwater reptiles, and mammals have been mainly considered. Conclusions so far indicate that:

(1) Infection of freshwater fishes occurs widely in the Holarctic region,

but is rare in the southern continents, where freshwater fish are more likely to share their ascaridoids with crocodilians.

(2) Amphibians have been found infected only in the warm parts of the Old World, reaching only as far north as the Mediterranean coast of Spain and as far east as New Guinea. They reached Australia in one frog species at the tip of Cape York Peninsula, but have evidently not reached Madagascar or any of the continental or volcanic islands.

(3) Among the reptiles, they occur in snakes in all warm regions of the world and there are many genera and species. The same is true of ascaridoids in crocodilians, though the two groups are only distantly related. Among other reptiles ascaridoids are sporadic and restricted in their range.

(4) Ascaridoids in terrestrial mammals are mainly found in Carnivora, among which an extensive radiation has occured with several genera and many species. These are mostly to be found in the Holarctic, Ethiopian and Oriental regions; endemic genera are almost entirely absent from the Neotropical and Australian regions.

The above is an over-simplification of the scanty data available on geographical distribution of ascaridoids, but what there is suggests strongly that the radiation and dispersal of the ascaridoids of terrestrial and freshwater hosts has occurred mainly after the breakup of the southern continents. The group as a whole has sparse representation on continental islands, with few endemic species. There is only one endemic genus in the Neotropical region and one in the Australian region, and the group is virtually absent in Oceania. The overwhelming majority of ascaridoids occur in the Holarctic, Ethiopian and Oriental regions.

From what is known of the affinities and evolution of ascaridoids of marine hosts it seems likely that they are derived from several different ascaridoid stems, rather than comprising a basic primitive stem.

It is perhaps too early to speculate on the evolution of ascaridoid nematodes, but there is little evidence to support Osche's hypothesis of parallel evolution or of their origin in sharks.

In my view the crocodilians are the most likely ancestral hosts and the estuaries of the great rivers of the Old World tropics the most likely centre of dispersal. Possibly the ascaridoids remained confined among crocodilians and a few other fluviatile hosts until the Tertiary. The emergence of the modern frogs possibly provided the first group of new hosts. But it seems likely that instead of radiating among the frogs, they took an evolutionary short cut (host-succession-extention) whereby they transferred from frogs to frog-eating reptiles. With the emergence of rodents they may have transferred from the rodent prey to the raptorial birds and terrestrial carnivores. Similarly, host-succession-extension appears to have occurred in the marine environment i.e. from fish to the fish-eating birds and mammals.

I see that I have come to the end of my time and, as our erstwhile vice-chancellor Sir Fred Schonell used to say — the longer the spoke, the greater the tire. So at the expense of a somewhat abrupt ending, I would like to thank you for so patiently listening to these ramblings.

Epilogue

> *"Time goes, you say? Ah no!*
> *Alas, Time stays,* we *go."*

It is four years since I gave this address and since then I have retired from my university position. My successor, Professor Colin Dobson, has honoured me by seeking my permission to publish it. I have gladly acceded, not because it has any literary merit, but because it allows me the opportunity to express in this epilogue my sincere thanks to all those who over the years have helped me in so many ways. Each of us garners a gallery of memory pictures of those who have counselled, cooperated, supported, encouraged, befriended, or loved, as we navigate our short journey down the refreshing river of Time. It is to these that I direct my thoughts and gratitude.

Part Two

Parasites

1 The Pterastericolidae: Parasitic Turbellarians from Starfish

L.R.G. Cannon

Introduction

Karling (1970) suggested that the origins of the Monogenea lie in the Pterastericolidae, a small group of parasitic turbellaria found in starfish. The first member of the group was described by Beklemischev (1916) from the gut of three species of *Pteraster* from the Gulf of Kola in the Russian arctic. He linked the worm with dalyellioid rhabdocoels notably *Graffilla* and *Paravortex*, parasites of molluscs. Westblad (1926) considered this worm belonged in the Dalyelliidae, but Meixner (1926) proposed a separate family Pterastericolidae closer to the Anoplodiidae (i.e., Umagillidae) than to the Graffillidae. Bresslau (1933) adopted this, though Stunkard and Corliss (1951) suggested that the worm, *Pterastericola fedotovi*, be the type of a subfamily (Pterastericolinae of the Umagillidae, the latter a group largely parasitic in echinoderms [Cannon 1982]). The discovery in starfish by Bashiruddin and Karling (1970) of *Triloborhynchus astropectinis* which, although superficially very different, was clearly related to *Pterastericola* confirmed the correctness of Meixner's view. Two further species, one in each genus, have since been recognized. This account records three more new species, summarizes the knowledge of the group and addresses the problem of the origins of the parasitic platyhelminths from their rhabdocoel relatives.

Materials and Methods

Starfish were collected by hand from the intertidal zone or by trawling, were returned alive to the laboratory and immediately dissected. It was found effectual to gently squash hepatic caeca between the upturned lid of a Petri dish and its base and examine the tissues using transmitted light with a stereoscopic microscope. Worms were evident as small pink spheres within the caecal pockets. Washings of the body cavity were also examined. Fixation for whole mounts was best in cold F.A.A. and for sections in Bouin's fluid followed by haematoxylin and eosin or Papanicolou's stain. In one case worms were fixed in Flemming's fluid which reveals lipid inclusions. Where appropriate (W) refers to specimens as whole mounts, (S) to serial sections and QM to collections in the Queensland Museum. All worms were collected by the author and measurements are derived from camera lucida drawings. Worms are drawn from the ventral surface.

Pterastericola vivipara

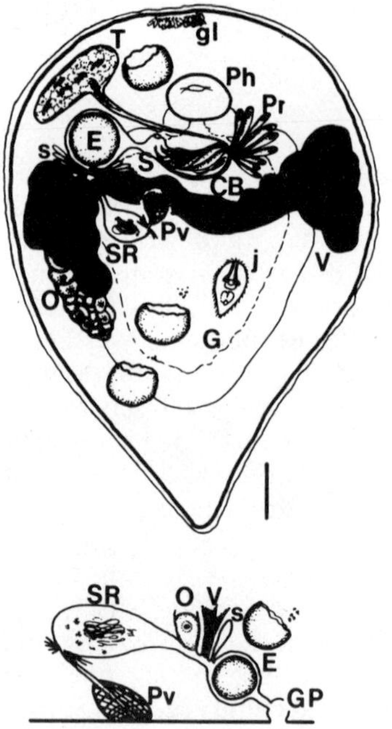

Figure 1.1 *Pterastericola vivipara* (scale = 100 μm) and a diagram of the female reproductive system. For abbreviations see p.31.

This species was reported by Cannon (1978) from *Acanthaster planci*, the Crown-of-Thorns starfish. Re-examination of the original material together with additional specimens shows the worms to be larger than reported earlier viz. 890 × 650 μm (640–1425 μm × 425–940 μm; $N = 20$) and further, the scale on the original figure should be 200 μm (not 250 μm). Examination of another worm from *Anthenea acuta* showed it to be indistinguishable from specimens from *A. planci*. Of particular interest is the pseudovagina. This clearly shows a break in the basement membrane and what appears to be a narrow fissure in the epithelial cells which form a plug about this break (fig.1.2,1.3). This fissure is essentially similar to that illustrated by Karling (1970) in *P. fedotovi*. It was also possible to follow proximally a narrow muscular duct (fig.1.4) from the pseudovagina. It opens into a seminal receptacle through a poorly developed sphincter (fig.1.5). Although Karling (1970) observed no kind

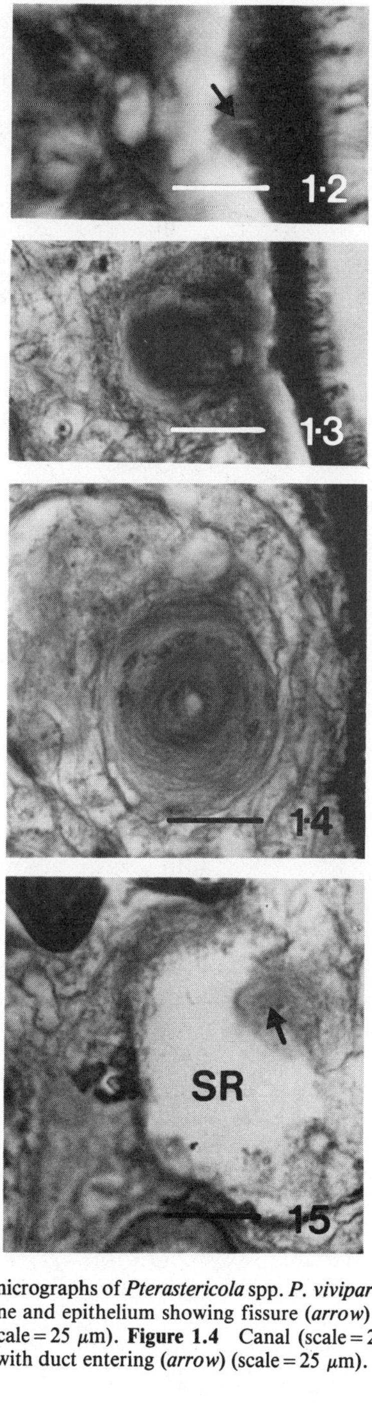

Figures 1.2–1.5 Photomicrographs of *Pterastericola* spp. *P. vivipara*, pseudovagina. **Figure 1.2** Basement membrane and epithelium showing fissure (*arrow*) (scale = 10 μm). **Figure 1.3** Below epithelium (scale = 25 μm). **Figure 1.4** Canal (scale = 25 μm). **Figure 1.5** At seminal receptacle (*SR*) with duct entering (*arrow*) (scale = 25 μm).

of communication existing between the pseudovagina and seminal receptacle of *P. fedotovi*, Beklemischev (1916) in his original description illustrates a short duct and states "un petit canal formé par un enforcement des teguments exterieur". I believe the development of the duct between seminal receptacle and its opening to the exterior may change with the age and state of sexual maturity of the worm (Cannon 1978).

Material examined

QM G10313-5 (W), QM G10316 (S), QM GL1585-1596 (W) ex *Acanthaster planci*, Centipede Reef, 9 June 1974. QM GL1582-4 (W) ex *Acanthaster planci*, Centipede Reef, 14 August 1974. QM GL1597 (S) ex *Anthenea acuta*, Moreton Bay, 1973.

Pterastericola sprenti n.sp.

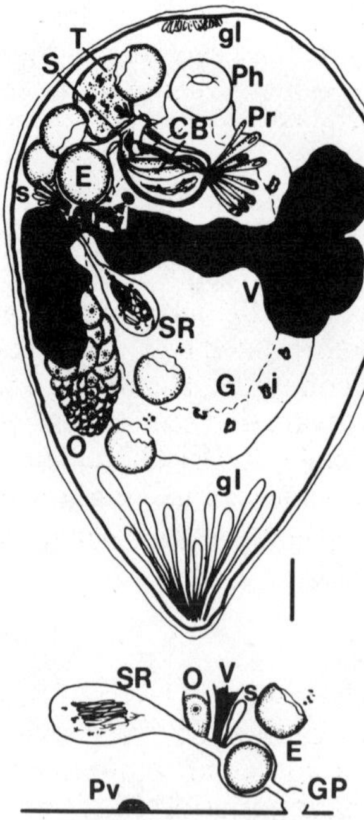

Figure 1.6 *Pterastericola sprenti* n.sp. (scale = 100 μm) and a diagram of the female reproductive system. For abbreviations see p.31.

Description

Small pyriform worms 976 × 623 μm (761–1232 μm × 558–725 μm, *N* = 8) red or pink, tapering at posterior, anterior tip mobile. Worms in hepatic caecal pockets of host, totally ciliated, epidermis more or less evenly thick 5–7 μm and cilia 3–4 μm, without rhabdites, occasional epidermal glands; normal sub-epidermal musculature, dorso-ventral muscles weak; parenchyma with scattered deeply staining cells, but especially with cluster of elongate, club-shaped, glandular cells opening at posterior tip of body. Mouth midventral about 1/8 from anterior end, pharynx doliiform, subspherical moderately well developed about 105 μm in diameter, crop short, large sac-like gut with tall columnar cells containing food vacuoles and notably numerous deeply staining inclusions (holocrine glands?) (fig.1.7). Testis single, weak walled sac in anterior right sector of body 143 × 100 μm (108–180 μm × 80–116 μm) with seminal duct crossing to centre body to enter bean-shaped copulatory bulb 145 × 75 μm (125–172 μm × 58–90 μm) at rear together with numerous club-shaped prostatic glands. Sperms and entwined prostatic cells separated within the muscular bulb, all discharge through a hollow male stylet — in this genus always coupled with a solid accessory hook (fig.1.8). Overall stylet and hook each about 50 μm long — the terminal portions about 25–30 μm long, stylet and accessory hook project into a common genital atrium and thence into a common gonopore just to right and behind pharynx. Ovary single, dextral 335 × 120 μm (265–400 μm × 80–220 μm) discharging at confluence of paired vitelline ducts and seminal receptacle in right sector of body about 1/3 from anterior end. Vitellaria form a roughly lobed band at this level extending a little posteriorly laterally on each side. Seminal receptacle a weak walled, blind vesicle 125 × 85 μm (80–145 μm × 55–145 μm) extending to midventral region. Pseudovagina (fig.1.9) represented by a small mass of muscles below the epidermis posterior to gonopore. Numerous shell glands discharge at confluence of vitellaria, ovary and seminal receptacle; ootype lined with epithelium, contains a single, spherical, golden yellow egg about 80 μm in diameter (72–94 μm), short uterine duct opens into common genital atrium, numerous egg capsules, some evidently hatched, lie scattered in parenchyma.

Diagnosis

Anterior gonopore, vitellaria more or less confined to band across anterior midbody. Numerous egg capsules in parenchyma of mature specimens, pseudovagina merely a muscular lump below ventral epithelium, numerous glands opening at posterior tip of body.

Details

Host: *Ophidiaster granifer* (Ophidiasteridae)
Location in host: caecal pockets
Locality: Heron Island, Great Barrier Reef, Australia
Collector: L. Cannon
Holotype: QM GL1600 (W), 10 December 1981
Paratypes: QM GL1601-5 (W), QM GL1606–7 (S), 10 December 1981

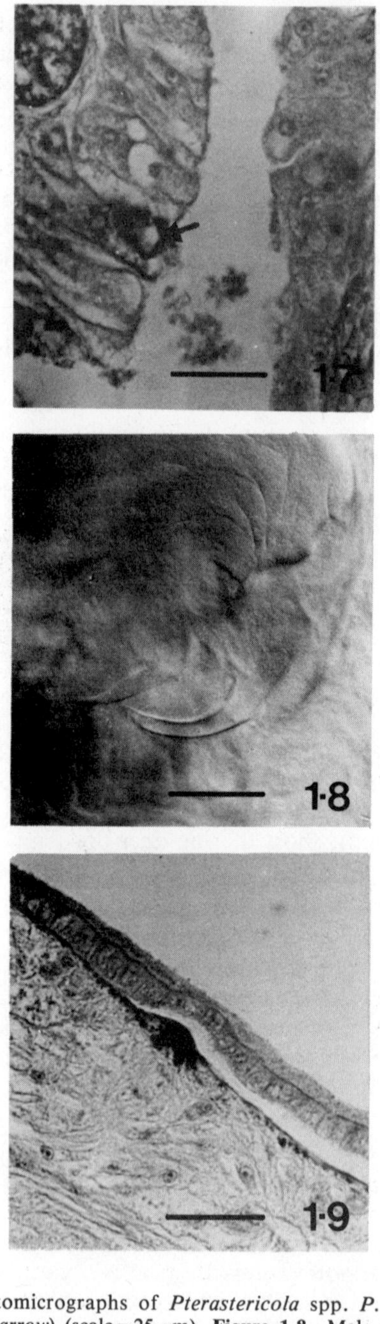

Figures 1.7–1.9 Photomicrographs of *Pterastericola* spp. *P. sprenti* n.sp. **Figure 1.7** Inclusion in gut (*arrow*) (scale = 25 μm). **Figure 1.8** Male stylets (scale = 25 μm). **Figure 1.9** Pseudovagina (scale = 25 μm).

Other material: QM GL1608-11 (W), 2 April 1974; QM GL1612-15 (S), 10 December 1981. Includes living specimens gently flattened under coverslip pressure.

Remarks

This species is distinguished from *P. vivipara*, its closest relative, by having a terminal cluster of glands and a more rudimentary pseudovagina. Both of these species differ from *P. fedotovi* since, unlike that species, the gonopore is in the anterior body and the vitellaria extend posteriorly.

Pterastericola australis n.sp.

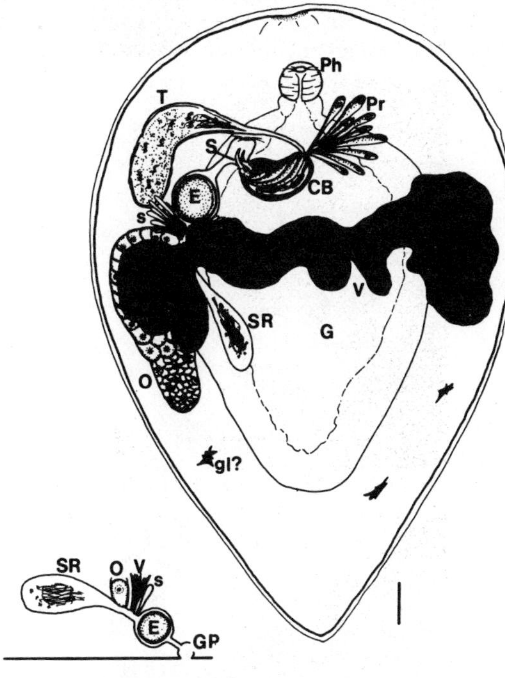

Figure 1.10 *Pterastericola australis* n.sp. (scale = 100 μm) and a diagram of the female reproductive system. For abbreviations see p.31.

Description

Small, pyriform, pink worms 1485 × 1037 μm (979–2011 μm × 747–1352 μm, $N=7$) found often contracted to subspherical shape within pockets in hepatic caeca of host. Ciliation uniform (6–8 μm) long, epithelium more or less uniformly thick (10–12 μm), without rhabdites, without conspicuous glandular or sensory apparatus (although the anterior tip of body may be quite mobile suggesting a sensory pit). Body masculature moderately weak, parenchyma has numerous deeply staining glandular? cells. Mouth midventral about 1/8 from anterior end, pharynx spherical, rather weak (about 95 μm in diameter) (fig.1.11), crop very short, dorsal sac-like gut with large tall columnar cells filled with lipid (fig.1.12). Testis single, thin walled sac 304 × 102 μm (162–445 μm × 54–178 μm) in anterior right sector of body at or just posterior to pharynx. Wide sperm duct enters back of copulatory bulb in midbody together with numerous, prominent prostatic glands, bulb thick walled, bean shaped, 158 × 91 μm (118–180 μm × 72–111 μm), about one third is seminal vesicle, remainder entwined prostatic glands. Bulb armed with solid accessory hook and hollow stylet which curve together, both about 36 μm long. Sperm and prostatic secretions discharge via stylet into common atrium and thence to gonopore which opens ventrally just behind and to right of mouth. Single, lobed, cylindrical ovary 463 × 135 μm (200–

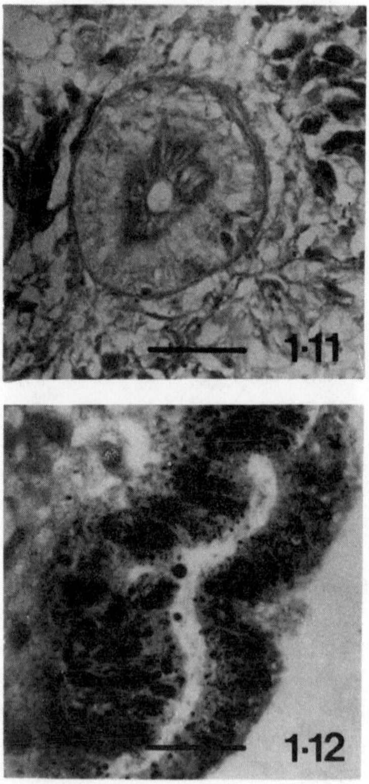

Figures 1.11–1.12 Photomicrographs of *Pterastericola* spp. *P. australis* n.sp. **Figure 1.11** Pharynx (scale = 50 μm). **Figure 1.12** Gut with lipid inclusions after fixation in Flemming's fluid (scale = 100 μm).

712 μm × 54–180 μm), in midlateral body behind testis. Confluence of paired vitelline ducts, seminal receptacle and ovary lies in midlateral body. Vitellaria lobed irregularly, one arm more or less ventral to ovary, other crosses body to left side and tends to extend slightly posteriorly, both more or less confined to anterior half of body. Seminal receptacle a blind sac 210 × 63 μm (126–414 μm × 45–90 μm) running posteriorly from confluence towards midventral body. Pseudovagina absent. Distally from confluence are numerous shell glands, a round ootype (epithelial lined and containing a single, subspherical golden yellow egg 110–120 μm in diameter); uterine duct short (100 × 30 μm), opens into common atrium.

Diagnosis

Anterior gonopore, vitellaria more or less confined to band across anterior midbody, single egg in ootype and none in parenchyma, no evidence of pseudovagina, pharynx rather weak.

Details

Host: *Patiriella calcar* (Asterinidae)
Location in host: caecal pockets
Locality: Hastings Point, NSW
Collector: L. Cannon
Holotype: QM GL1616 (W), June 1974
Paratypes: QM GL1617 (W), 12 March 1980; QM GL1618 (W) 22 August 1973; QM GL1619–20 (W), 30 April 1974; QM GL1621 (S), 11 March 1980; QM GL1622 (S), 20 July 1982
Other material:QM GL1623 (W), 20 July 1982; QM GL1624–5 (W), 3 March 1974; QM GL1626–9 (S), 3 March 1980; QM GL1630 (S), 20 July 1982; QM GL1631 (S), 21 May 1974; and living material from *Patiriella calcar* from Hastings Point.

Remarks

This species is distinguished from other species of *Pterastericola* in completely lacking any semblance of a pseudovagina. It is somewhat larger than the other Australian species to which it bears a superficial resemblance, but has a weak pharynx and only one egg, none in the parenchyma.

Pterastericola ramosa n.sp.

Description

Small, pyriform, pink worm, ciliated all over measures 1232 × 688 μm and 1522 × 725 μm (holotype and paratype respectively) found in hepatic caeca of host. Epidermis uniformly thick without conspicuous glands or sensillae, though a cluster of parenchymal glands discharge at anterior-

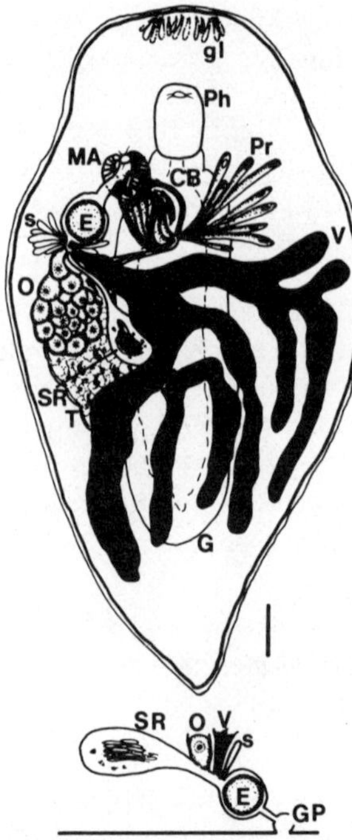

Figure 1.13 *Pterastericola ramosa* n.sp. (scale = 100 μm) and a diagram of the female reproductive system. For abbreviations see p.31.

ventral tip. Mouth at about 1/8 from anterior end, midventral, pharynx slightly tubular (116 × 87 μm and 145 × 110 μm) followed by a short crop about 90 μm in diameter and a long moderately tubular dorsal gut stretching to posterior quarter of body. Single, weak walled, lobed, testis (300 × 200 μm and 300 × 100 μm) along midlateral, right sector of body. A broad sperm duct runs ventral to gut at level of ootype and enters copulatory bulb in midbody; numerous, club-shaped prostatic glands in anterior left sector of body. Bulb thick walled, muscular and subspherical (125 μm and 110 μm in diameter) containing entwined ducts of prostatic glands and seminal vesicle, distally bulb armed by accessory hook and stylet (fig.1.14) each with base anchored to bulb with muscle insertions. Overall (including basal part) the hook is shorter (100 and 80 μm long) than stylet (120 and 140 μm long), each about 5–6 μm in diameter. Hook and stylet appear to oppose one another like index finger and thumb. These structures open to a muscular male antrum about 90 μm in diameter (with walls 30 μm thick) (fig.1.15) and thence to common atrium and common gonopore ventral and a little behind and to left of mouth. Single, lobed ovary about 150 μm in diameter, ventral and partly anterior to testis, opens directly to confluence of vitelline ducts and seminal receptacle

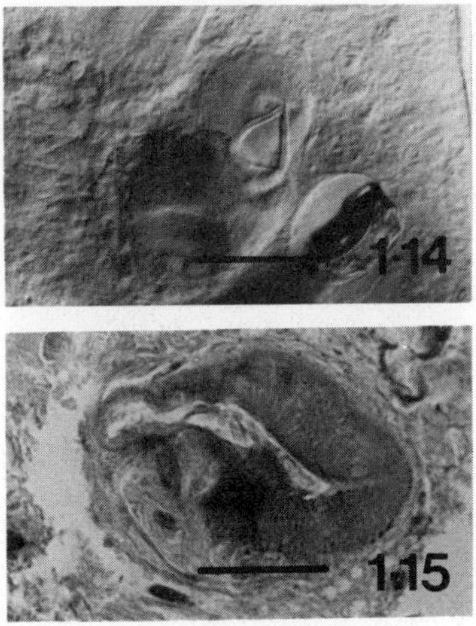

Figures 1.14–1.15 Photomicrographs of *Pterastericola* spp. *P. ramosa* n.sp.
Figure 1.14 Male copulatory stylets (scale = 100 μm).
Figure 1.15 Muscular male antrum (scale = 50 μm).

together with numerous shell glands discharging into ootype lined with epithelium and containing a single, round, golden yellow egg (85 μm and 75 μm). Short narrow uterine duct from ootype to common atrium. Seminal receptacle thin walled vesicle full of sperm and debris about 200 × 150 μm in ventral midbody, duct runs anteriorly and dorsally to join confluence. Pseudovagina absent. Vitellaria in posterior two-thirds of body, branching with five or six long arms (500–700 μm long and 50–60 μm wide) running across and posteriorly close to ventral surface.

Diagnosis

Anterior gonopore, only single egg in ootype and none in parenchyma, pseudovagina absent, vitellaria branched, copulatory bulb subspherical, stylet and hook of male cuticular apparatus opposed, muscular male antrum, testis in midlateral body extending posterior to ovary, pharynx slightly tubular, not subspherical.

Details

Host: *Luidia australiae* (Luidiidae)
Location in host: caecal pockets
Locality: Moreton Bay, Queensland
Collector: L. Cannon
Holotype: QM GL1598 (W), 16 August 1973
Paratype: QM GL1599 (S), 16 August 1973

Remarks

The ramifying and posteriorly directed vitellaria, the opposed stylet and accessory hook and the muscular male antrum serve to distinguish this species from all others of the genus. Furthermore, its pharynx, though well developed, is slightly tubular rather than subspherical.

Discussion

Internally there is a fair degree of homogeneity among the Pterastericolidae — single testis, muscular copulatory bulb, stylet with accessory hook as copulatory structures, the single ovary, paired vitellaria, ootype without uterus, common midventral gonopore and a seminal receptacle (sometimes referred to as a bursa). Perhaps the most interesting feature is the degree to which a true or pseudovagina is developed. Beklemischev (1916) illustrated a short canal from the seminal receptacle to the body wall surrounded by a muscular sphincter, but no pore. Karling (1970) re-examined *P. fedotovi* and found, in one specimen, a slit in the epidermis and a pit in the basement membrane, but no pore. In others he found no evidence of a pore, nor indeed any evidence of a connection between seminal receptacle (bursa) and the exterior in any specimens examined. Cannon (1978) reported an apparent pore in one young specimen of *P. vivipara*, but it was not evident in older specimens. A small, muscular lump below the ventral epidermis was recognized as a pseudovagina in *Triloborhynchus astropectinis* by Bashiruddin and Karling (1970); no connection was seen between this and the seminal receptacle. A similar muscle lump is recognized as a pseudovagina in *P. sprenti* described herein. In *T. psilastericola* the presence of such a pseudovagina was not recorded; however, no difference from the situation in *T. astropectinis* was noted by Jespersen and Lützen (1972). Since the small muscular lump is an easily overlooked characteristic its presence in *T. psilastericola* needs confirmation. A pseudovagina is definitely absent in material of the new species, *P. australis* and *P. ramosa*, from Australia. All the evidence suggests that (1) the vagina may be functional in young or protandric specimens and atrophy later and (2) that apart from changes during an individual's life, there has been an apparent evolutionary loss of this structure so that the muscular lump seen in *P. sprenti* and *T. astropectinis* lies intermediate between the states found in *P. vivipara* on one hand and *P. australis* on the other.

A similar transition appears to have taken place in the position of the gonopore and the distribution of the vitellaria. The gonopore is well posterior in *P. fedotovi* but behind the mouth in *Triloborhynchus*, a clearly more derived group. The gonopore is also near the mouth in the species from Australia. It seems better to alter the generic definition given by Bashiruddin and Karling (1970) by deleting mention of gonopore position and to accept these Australian species as members of *Pterastericola* because of their similarity to *P. fedotovi* than to create a new genus. It appears, therefore, that there has been an evolutionary anterior migration of the gonopore; the vitellaria have firstly become a more or less transverse band and then, in the case of *P. ramosa*, have ramified posteriorly.

Another transition appears to be the progressive loss of ciliation. Jespersen and Lützen (1972) reported a loss of ciliation from young to older animals. Similar reductions between young and old *Cleistogamia* (Umagillidae) were reported by Cannon (1982). Evidently the restricted ciliation in *Triloborhynchus* is derived and perhaps parallels adoption of a more intimate symbiotic (parasitic) form of association. Perhaps the accummulation of eggs (or the production of a larger number of eggs) is also associated with a more parastic lifestyle in *T. psilastericola, P. vivipara* and *P. sprenti*.

The male apparatus of *Triloborhynchus* appears to be derived from one resembling that of the dalyelliid, *Microdalyellia*, which has horns, scoop and spine (see Luther 1955, fig.20), although bearing a strong resemblance to the hook and stylet of the *Danorhynchus-Scanorhynchus-Neopolycystis* groups of the kalyptorhynch family Polycystididae (see Karling 1955). Presumably the horns of the apparatus have been lost in *Pterastericola*.

The relationships among the Peterastericolidae are presented as a cladogram in figure 1.16.

Although considered sufficiently different from other rhabdocoel turbellarians to be placed in its own family by Hickman and Olsen (1955), the aberrant starfish parasite *Acholades asterias* may arguably be placed near the Pterastericolidae. Hickman and Olsen (1955) argued that the nerve centre was posterior near the gonopore presumably because the gonopore is terminal in umagillids to which they considered *Acholades* had some affinities. Equally this nerve centre could be anterior with an anterior sensory pit and an anterior gonopore. An anterior sensory region and a gonopore which shows evolutionary anterior movement are seen in the Pterastericolidae. An anterior movement of the musculature of the pseudovagina so as to be incorporated into an anterior muscular bursa would result in a structure similar to the muscular bursa associated with the gonopore (together with an indeterminate canal from it to the parenchyma) reported for *Acholades*. Loss of alimentary structures was noted by Hickman and Olsen (1955) to be a feature of the Fecampiidae; but similar loss is seen also in the umagillid *Fallacohospes inchoatus*. Until more is known the Acholididae will remain an enigma, but it seems likely it is a specialized endoparasitic derivative of the Pterastericolidae.

The worms of the F. Pterastericolidae have been found only in starfish from around the North Sea and Arctic and from the east coast of

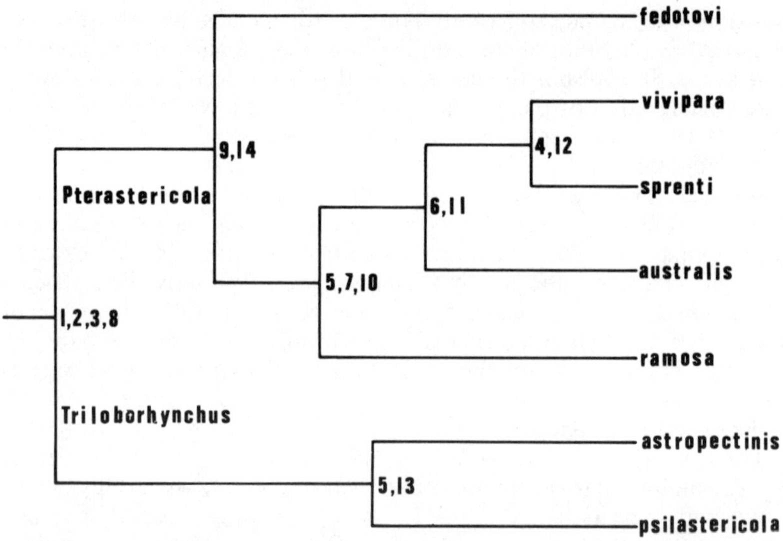

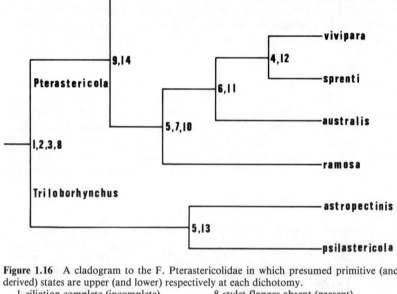

Figure 1.16 A cladogram to the F. Pterastericolidae in which presumed primitive (and derived) states are upper (and lower) respectively at each dichotomy.

1 ciliation complete (incomplete)
2 anterior adhesive organs/pits absent (present)
3 posterior adhesive organs absent (present)
4 posterior glands absent (present)
5 pharynx sub-spherical (tubular)
6 pharynx strong (weak)
7 male antrum weak (muscular)
8 stylet flanges absent (present)
9 vitellaria extend anterior (posterior)
10 vitellaria unbranched (branched)
11 pseudovagina present (absent)
12 pseudovagina well developed (remnant)
13 single egg (many eggs)
14 genital pore in posterior body (in anterior body)

Australia (table 1.1). This almost certainly reflects not so much their occurrence as where they have been sought. There appears to be little significance in the host family relationships. Although both species of *Triloborhynchus* are from Astropectinidae, *P. vivipara* occurs in starfish in families as diverse as Acanthasteridae and Goniasteridae. Although clearly symbiotic there has been some discussion in the past whether these worms were parasitic. For example, *T. astropectinis* was considered by Bashiruddin and Karling (1970) to feed predominantly on caecal contents rather than directly on the tissues, but *T. psilastericola* was shown by Jespersen and Lützen (1972) to be a true parasite. Cannon (1978) reported that *P. vivipara* fed on the lipid rich, hepatic caecal epithelium of its host. In fact it accumulates lipid which is rapidly depleted upon starvation in its own intestine. *P. australis* also accumulates lipid in gut cells (fig.1.12), but the parenchyma is rich in stored glycogen. According to Jennings (1977) endoparasites predominantly rely on glycogen, whereas free-living animals use more lipid. The pterastericolids appear to reflect nutritionally as well as behaviourally and morphologically specializations characteristic of the great bulk of the parasitic platyhelminths.

Table 1.1 Check list of the Pterastericolidae

Species	Host (family)	Locality	Authority
Pterastericola fedotovi	*Pteraster pulvillus* (Pterasteridae)	Gulf of Kola, White Sea	Beklemischev (1916)
	P. obscurus (Pterasteridae)	Gulf of Kola, White Sea	Beklemischev (1916)
	P. militaris (Pterasteridae)	Gulf of Kola, White Sea	Beklemischev (1916)
	P. militaris (Pterasteridae)	Kandalaksha Bay, White Sea	Karling (1970)
P. vivipara	*Acanthaster planci* (Acanthasteridae)	Great Barrier Reef	Cannon (1978)
	Anthenea acuta (Goniasteridae)	Moreton Bay, Qld	Present account
P. sprenti	*Ophidiaster granifer* (Ophidiasteridae)	Heron Island, Great Barrier Reef	Present account
P. australis	*Patiriella calcar* (Asterinidae)	Hastings Point, NSW	Present account
P. ramosa	*Luidia australiae* (Luidiidae)	Moreton Bay, Qld	Present account
Triloborhynchus astropectinis	*Astropecten irregularis* (Astropectinidae)	Cullercoats, Plymouth, Great Britain	Bashiruddin & Karling (1970)
		Kristineberg, Sweden	Bashiruddin & Karling (1970)
		Korsfjorden, Norway	Bashiruddin & Karling (1970)
T. psilastericola	*Psilaster andromeda* (Astropectinidae)	Oslo Fjord, Norway	Jespersen & Lützen (1972)

Much has been written on the evolution of the various parasitic flatworm groups — e.g. Monogenea (Llewellyn 1965); Digenea (Pearson 1972); Cestoda (Freeman 1973) — and all agree with the sentiments of Bresslau (1933) that the groups arose from a rhabdocoel-like ancestor. Although direct lineages are no doubt unrealistic there are charactistics and qualities among the rhabdocoels, especially within the suborder Dalyellioida which illustrate possible ancestral traits. It is noteworthy that symbiotic members are found in all families of the Dalyellioida (although the Dalyelliidae and Provorticidae are predominantly free-living in fresh and salt water respectively).

I believe four lifestyle trends can be recognized in the Dalyellioida: (1) endosymbiotic typified by parenteral development (Fecampiidae); (2) endosymbiotic with enteral and parenteral development (Umagillidae, Pterastericolidae, Acholadidae); (3) free-living and ectosymbiotic (Dalyelliidae); and (4) free-living and endosymbiotic in spaces giving access to the exterior (gut, gonads, kidneys) (Graffillidae). The fecampiids are elongate worms which have lost all trace of an alimentary system. This together with their early parental development followed by egg laying on emergence strongly suggests the primitive cestode proposed by Freeman (1973). The graffillids live both free and as endosymbionts of molluscs. In both groups juveniles may develop in the parent body. (Similar young in *Pterastericola vivipara*, for example, have anterior glands and must penetrate the parent's tissues, at least.) Such characteristics as having juveniles capable of tissue penetration and development of young within the parent body together with a predisposition for molluscs are clearly indicative of early digeneans. The umagillid-pterastericolid group have radiated in echinoderms and perhaps are a terminal group, although Karling (1970) linked the pterastericolid *Triloborhynchus* with the Udonellida — a primitive monogenean group according to van der Land (1967). Both Ivanov (1952) and Nichols (1975), however, consider this group much closer to the Temnocephalida than the Monogenea. Williams (1981) quotes Matjasic as believing the Monogenea were derived from a temnocephalid group. Although Williams (1981) favoured a separate class for the Temnocephalida, following Baer (1961), she did recognize their close resemblance to *Dalyellia*. Karling (1970) thought the Temnocephalida closer to the Dalyelliidae than to the Pterastericolidae and Ax (1963) also saw close ties between Temnocephalida and, at least, the Dalyellioida. Regardless

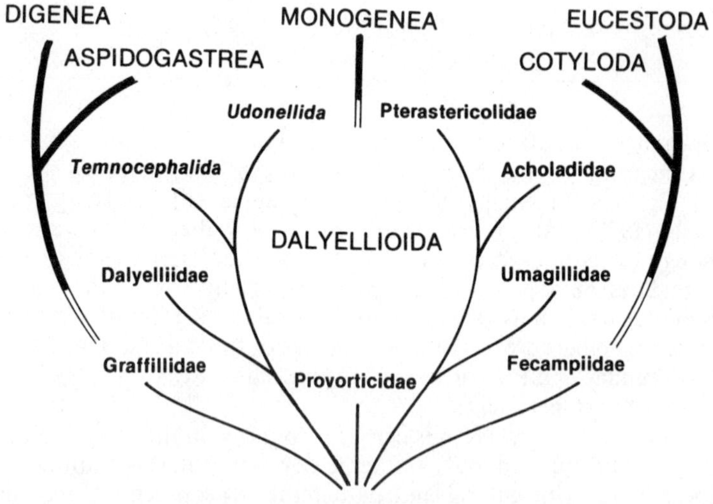

Figure 1.17 Synopsis of relationships among the Dalyellioida (O. Rhabdocoela) and their probable position in the evolution of the parasitic flatworms.

of the exact taxonomic status of the Temnocephalida and Udonellida there do seem to be similarities between them and the Monogenea which are reminiscent of the Dalyelliidae rather than the Pterastericolidae.

Figure 1.17 shows some possible relationships within the Dalyellioida and to the groups Temnocephalida and Udonellida. The origins of the parastic flatworms are obscure, but lifestyle parallels to those of these forms do occur in the Dalyellioida. Perhaps investigation of the dalyellioid groups will throw light on the behavioural, physiological and morphological specializations of the antecedants of today's parasitic flatworms.

Acknowledgments

For practical assistance in obtaining material or assistance in the field or laboratory I wish to thank Dick Martin, Con Boel, Gerry Goeden and Neal Hall, but for an opportunity to study marine parasitology and encouragement to consider matters philosophical I am indebted to J.F.A. Sprent and to others he has influenced.

Abbreviations used in figures

CB	= Copulatory bulb	*O*	= ovary
E	= egg	*Ph*	= pharynx
G	= gut	*Pr*	= prostate glands
gl	= gland	*Pv*	= pseudovagina
GP	= gonopore	*S*	= stylets
i	= inclusion	*s*	= shell glands
j	= juvenile	*SR*	= seminal receptacle
MA	= muscular antrum	*T*	= testis

V = vitellaria

References

Ax, P. 1963. Relationships and phylogeny of the Turbellaria. In *The lower metazoa*, ed. E.C. Dougherty, 191–224. Berkeley: University of California Press.

Baer, J.G. 1961. Classe des temnocephales. In *Traité de Zoologie*, ed. P.P. Grassé, **4**:213–41. Paris: Masson et Cie.

Bashiruddin, M. and Karling, T.G. 1970. A new entocommensal turbellarian (Fam. Pterastericolidae) from the sea star *Astropecten irregularis. Zeitschrift für Morphologie und Ökologie der Tiere* **67**:16–28.

Beklemischev, V. 1916. Sur les turbellairés parasites de la côte Mourmanne. II.Rhabdocoela. *Trudȳ Imperatorskago S. — Peterburgskago obshchestva estestvoispȳtateleť* **45**:1–78 (French summary pp. 60–73).

Bresslau, E. 1933. Turbellaria. In *Handbuch der Zoologie*, ed. W. Kükenthal and T. Krumback, **2**:52–320. Berlin: Walter de Gruyter.

Cannon, L.R.G. 1978. *Pterastericola vivipara* n.sp., a parasitic turbellarian

(Rhabdocoela:Pterastericolidae) from the Crown-of-Thorns starfish, *Acanthaster planci*. *Memoirs of the Queensland Museum* **18**:179–83.

Cannon, L.R.G. 1982. Endosymbiotic Umagillids (Turbellaria) from holothurians of the Great Barrier Reef. *Zoologica Scripta* **11**:173–88.

Freeman, R.S. 1973. Ontogeny of cestodes and its bearing on their phylogeny and systematics. *Advances in Parasitology* **11**:481–557.

Hickman, V.V. and Olsen, A.M. 1955. A new turbellarian parasitic in the sea-star, *Coscinasterias calamaria* (Gray). *Papers and Proceedings of the Royal Society of Tasmania* **89**:55–63.

Ivanov, A.V. 1952. Morphology of *Udonella caligorum* Johnston, 1835, and the position of Udonellidae in the systematics of Platyhelminths. *Parasitological collection of the Institute of Zoology Academy of Science USSR* **14**:112–63 (English translation 25, Virginia Institute of Marine Science).

Jennings, J.B. 1977. Patterns of nutritional physiology in free-living and symbiotic turbellaria and their implications for the evolution of entoparasitism in the phylum Platyhelminthes. *Acta Zoologica Fennica* **154**:63–79.

Jespersen, A. and Lützen, J. 1972. *Triloborhynchus psilastericola* n.sp. a parasitic turbellarian (Fam. Pterastericolidae) from the starfish *Psilaster andromeda* (Müller and Truschel). *Zeitschrift für Morphologie und Ökologie der Tiere* **71**:290–98.

Karling, T.G. 1955. Studien über Kalyptorhynchien (Turbellaria). *Acta Zoologica Fennica* **88**:1–43.

Karling, T.G. 1970. On *Pterastericola fedotovi* (Turbellaria), commensal in sea stars. *Zeitschrift für Morphologie und Ökologie der Tiere* **67**:29–39.

Land, J. van der, 1967. Remarks on the subclass Udonellida (Monogenea) with a description of a new species. *Zoologische Mededlingen* **42**:69–81.

Llewellyn, J. 1965. The evolution of parasitic platyheminths. In *Evolution of parasites*, ed. A.R. Taylor, 47–78. Oxford: Blackwell Scientific.

Luther, A. 1955. Die Dalyelliiden. *Acta Zoologica Fennica* **87**:1–337.

Meixner, J. 1926. Beitrag zur Morphologie und zum System der Turbellaria — Rhabdocoela. II. Über *Typhlorhynchus nanus* Laidlaw und die parasitischen Rhabdocölen nebst Nachtragen zu den Calyptorhynchia. *Zeitschrift für wissenschaftliche Biologie* **5**:577–624.

Nichols, K.C. 1975. Observations on lesser known flatworms: *Udonella*. *International Journal for Parasitology* 5: 475-82.

Pearson, J.C. 1972. A phylogeny of life-cycle patterns of the Digenea. *Advances in Parasitology* **10**:153–89.

Stunkard, H.W. and Corliss, J.O. 1951. New species of *Syndesmis* and a revision of the family Umagillidae Wahl, 1910 (Turbellaria:Rhabdocoela). *Biological Bulletin. Marine Biological Laboratory, Woods Hole* **101**:319–34.

Westblad, E. 1926. Das Protonephridium der parasitischen Turbellarien. *Zoologischer Anzeiger* **67**:323–33.

Williams, J.B. 1981. Classification of the Temnocephaloidea (Platyhelminthes). *Journal of Natural History* **15**:277–99.

2 Field and Laboratory Observations on the Oyster Parasite *Marteilia sydneyi*

R.J.G. Lester

Introduction

The oyster industry in northern New South Wales and southern Queensland is periodically affected by epizootics that kill off most of the cultured and wild oysters in localized areas, particularly estuaries and enclosed bays. The mortalities occur primarily in the summer; because of this many oyster farmers leave their leases empty for January, February and March.

The oysters are killed by the ascetosporan protozoan *Marteilia sydneyi* first reported by Wolf (1972) from southern Queensland and later described by Perkins and Wolf (1976) from material collected in the same area. Field observations indicated that it takes less than 60 days for the oysters to die after they become infected (Wolf 1979).

Ascetosporans have been the cause of major mortalities in oyster stocks in the United States and France. Despite a concentrated research effort much of their biology, including their life cycle, remains unknown.

My attempts to experimentally infect *C. commercialis* with *M. sydneyi*, by exposing oysters to spores under a variety of conditions and by transplanting pieces of infected digestive gland were unsuccessful. Therefore, the following questions were addressed. Where and under what conditions did oysters become infected in the field? Were some strains of oysters refractory to infection? And what conditions were necessary for the disease to develop? In addition, six drugs were tested to ascertain whether they could kill the parasite or improve host survival.

Materials and Methods

Uninfected oysters *Crassostrea commercialis* were obtained from Port Stephens, 200 km south of the most southerly report of *Marteilia sydneyi*, the Macleay River. They were held in recirculating sea water tanks at 20°C or 25°C for up to 6 months before being used. Queensland oysters belonging to the same species were obtained from Tin Can Bay, Pummicestone Passage (Tripcony Bight), and Moreton Bay (Stradbroke Island). In field experiments, samples of about 25 oysters enclosed in nylon bags (50 mm stretched mesh) were placed on trays on three Queensland oyster leases, one at the Aldershots, 50 km south at Brisbane in the Southport Broadwater, and two (Tripcony Bight and Ningi Creek) in Pummicestone Passage, 50 km north of Brisbane. On four occasions paired samples of oysters, one from Port Stephens and one from southeast Queensland, were put in the field side by side.

Infected oysters were obtained from commercial growers at times of epizootics. Oysters were checked for the presence of spores by microscopic examination of a squash preparation of part of the digestive gland. At the same time a piece of the gland was fixed in Davidson's fixative for later sectioning and staining with haematoxylin and eosin. For the drug trials oysters were treated in three groups. In the first, oysters with sporulating infections were immersed in a solution of the drug in seawater for 2 days then returned to the lease for 3 weeks. In the second and third groups, oysters with infections in pre- and early-sporulation stages were given the drug by injection into the digestive gland using a 1.1 mm needle through a 1.5 mm hole drilled in the shell, and kept dry for the following 2 days. The second group was then returned to the lease for 5 weeks (in May/June after chance of further natural infection had passed), whereas the third group was given a second injection 3 days later and then returned to the lease for 4.5 weeks. The drugs used were: pyrimethamine and sulphadoxine (Fansidar^R; Roche, Sydney, NSW); trimethoprim and sulphamethoxazole (Bactrim^R; Roche); chloroquine phosphate (Chlorquin^R; Fisons, Glebe, NSW); pyrimethamine alone (Daraprim^R; Wellcome, Rosebury, NSW), all obtained in tablets which were broken up in seawater; and acriflavine solution (Rockes, Thebarton, SA).

Ningi Creek rainfall was estimated by averaging the figures for Bribie Island and Caboolture, each 12 km away in opposing directions. The temperature data are for Southport, 12 km south of the Aldershots. Creek flushing rate was estimated by floating a perforated object 300 x 200 x 60 mm vertically 100 mm below the surface. It was put in at high water at the furthest point up the creek that oysters grow and observed until low water. This was done on 18 April 1982 when there was a neap tide (tide difference 1.2 m).

Results

To examine the possible influence of genetic stock on the susceptibility of oysters to *M. sydneyi* in Queensland waters, paired samples of uninfected oysters (Queensland stock and NSW stock) were exposed to infection at three experimental field locations in southeast Queensland (table 2.1). In each case the samples were retrieved 4 to 6 weeks after a natural outbreak of QX disease. This gave sufficient time for spore production but was insufficient to cause mortality.

In each case prevalences were not significantly different ($P > 0.05$) indicating no difference between stocks in their susceptibility to infection. To determine if stocks showed any differential ability to recover from

Table 2.1 Comparison of susceptibility of oysters from different genetic stock (Queensland and NSW) to infection with *Marteilia sydneyi* at three field locations in southeast Queensland

Trial	Field site	Source (stock)	Total no.	No. with spores*	% with spores*
1	Ningi Creek	Pummicestone Passage, Qld	15	6	40
		Pt Stephens, NSW	15	6	40
2	Tripcony Bight	Pummicestone Passage, Qld	15	1	7
		Pt Stephens, NSW	15	1	7
3	Aldershots	Tin Can Bay, Qld	55	32	58
		Pt Stephens, NSW	51	31	58
4	Aldershots	Moreton Bay, Qld	17	14	82
		Pt Stephens, NSW	18	11	61

* Based on squash technique

infection, a 5th trial begun in parallel to trial no. 4 was allowed to continue for a further 3 months, by which time most mortalities would be expected to have occurred. Of 21 Queensland oysters, 16 or 76% had died, and 13 of 25 NSW oysters (52%) had died. These mortalities were not significantly different ($X^2 = 2.86$; $P > 0.05$).

To determine the influence of salinity and temperature on the course of the infection, 22 infected oysters were placed in each of four laboratory regimes, viz. 25°C and 30°C at 15 and 30°/00 salinities. In each case mortality was nearly total, no more than 3 oysters surviving more than 2 months in any of the groups. Over half the reserve stock kept at 20°C and 30°/00, however, survived for more than 2 months.

Preliminary experiments with implanting infected tissue failed to produce infections. This may have been because the transferred parasites had already completed their reproductive phase in the oyster. It was, therefore, necessary to find out when oysters were becoming infected in the field. Oysters were exposed in three field locations and for varying lengths of time (fig. 2.1). At the Aldershots infections were obtained from December through to early April. At Tripcony Bight new infections occurred during February and March. At Ningi Creek they occurred during a few days in January (coinciding with heavy rain, see Discussion), and during April, though no spores were produced in the second infection before the experiment terminated. Exposure at other times both earlier and later at all three locations did not result in infection. Evidently oysters became infected only at specified times and this varied with the locality.

In three samples, infections remained over the winter. At the Aldershots, two carried spores from March/April through to September (fig. 2.1; 0.6X, 0.7X). Samples after the first week of April had remained uninfected. In May, the monthly average for the daily maximum temperature fell below 25°C (fig. 2.1). The sample placed in Ningi Creek in early April suffered some mortality in late April and trophozoites were found in sections but no spores were produced. Trophozoites without

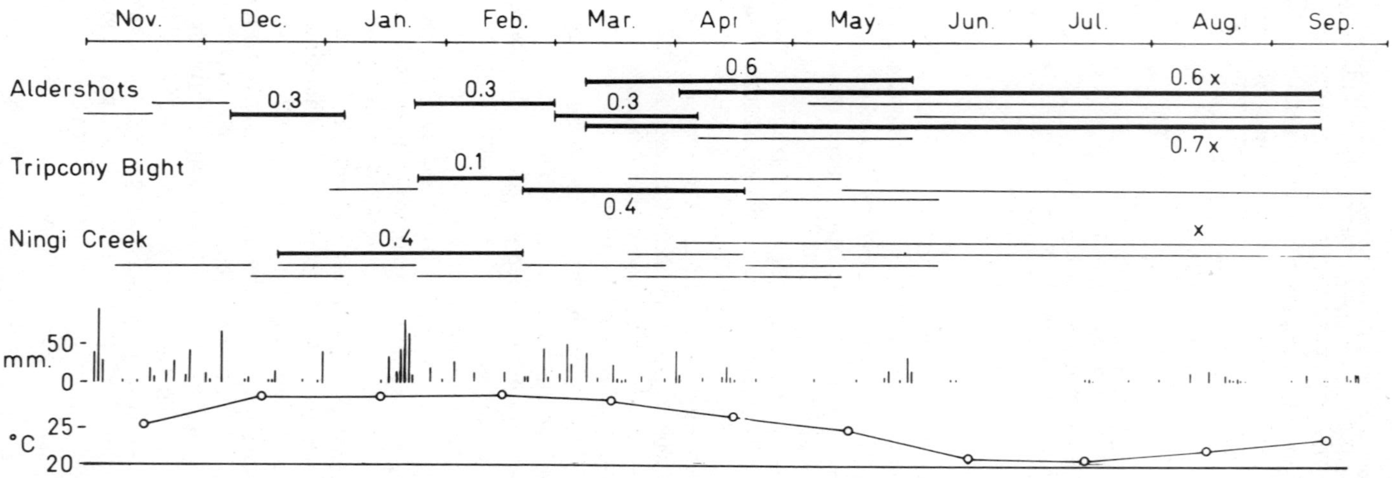

Figure 2.1 Periods for which test samples were placed on three commercial oyster leases in 1981–82. Each line represents one sample. Heavy lines indicate that spores were found when the sample was removed. The decimal gives the proportion of oysters positive. Samples marked *X* are those in which the parasite overwintered. Below these lines are shown the approximate daily rainfall for Ningi Creek, and the monthly averages for the daily maximum air temperature near the Aldershots (Southport).

spores were still present the following spring indicating that little development occurred during the winter. Sections of other overwintering samples showed that they had remained uninfected.

To determine the focus of infection in Pummicestone Passage samples of wild oysters were examined in March (fig. 2.2). Highest infection levels

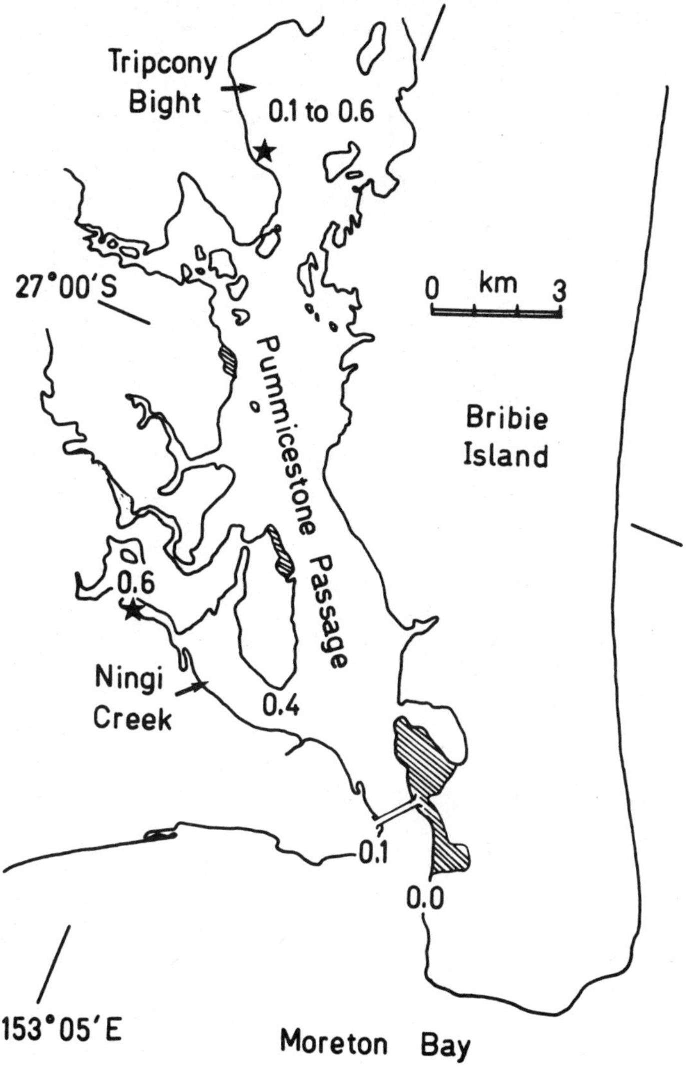

Figure 2.2 Southern half of Pummicestone Passage showing the proportion of wild oysters carrying spores in March 1982. Stars mark the location of the two regular test sites. Shading indicates residential areas.

were in upper Ningi Creek and in part of Tripcony Bight. This suggested that the Ningi Creek infection had originated in the creek and not come from outside. A buoy floated down the creek on a neap tide passed well out into Pummicestone Passage by low water, showing that the creek was well flushed with each tide. This may account for the short period over which the infective stage was present (see Discussion).

Of the drugs tested, none significantly improved the survival of infected oysters after the immersion treatment, and all but chloroquine were shown to fail to kill the parasite after inoculation (table 2.2). In histological sections the parasites appeared to be healthy in all infected survivors. Only one oyster survived the chloroquine trial; when fixed it was in poor condition but was uninfected.

Table 2.2 Survival of oysters infected with *Marteilia sydneyi* 3 to 5 weeks after one of three treatments with various drugs at concentrations of from 1 to 100 times the recommended human dose

Drug	Dose	Number surviving/number treated		
		Immersion treatment	Single injection	Double injection
Control		8/22	—	2/5*
Fansidar[R]	x1	4/23	—	—
	x10	—	0/5	1/5*
	x100	—	0/5	0/5
Bactrim[R]	x5	9/24	—	—
	x50	—	1/5*	1/5+
	x100	—	2/5*	1/5*
Daraprim[R]	x10	9/17	1/5*	0/5
	x100	—	0/5	0/5
Chlorquin[R]	x1	4/17	—	—
	x10	—	1/5+	0/5
Acriflavine mg/g oyster tissue	0.01	—	2/5*	1/5*

R Registered trademark
* Parasites found in sections of surviving oysters
\+ Surviving oyster uninfected

Discussion

The failure of the transplants to infect oysters was evidently not associated with any genetic resistance in the recipients as the results showed no significant difference between strains in their susceptibility to the disease. This is in agreement with Wolf (1979) who observed that wild stocks appeared to be just as vulnerable to invasion as cultivated stocks.

Both field and laboratory observations indicated that temperature affected the course of the disease, lower temperatures slowing development and retarding host mortality. Neither observation implicated salinity as being important once the oysters were infected.

Given a sufficiently high temperature for development, for transplants to produce the disease in another oyster it is probably necessary to transfer presporulation stages as once sporulation starts the parasite undergoes little further reproduction (Perkins & Wolf 1976). The presporulation stages have not been described. They are presumably in the connective tissue of the digestive gland as this is the site of a massive inflammatory response early in the infection.

The precise time of infection could be determined with reasonable accuracy for the first Ningi Creek outbreak. The 19 December to 20 February sample developed spores (fig. 2.1; 0.4). A concurrent sample from 25 January to 20 February remained uninfected so the infection must have occurred prior to 25 January. It was evidently only a short time before the 25th as a sample taken out on this date was free of spores, presumably because the parasite had not had time to develop, though the inflammatory response mentioned above was evident in sections. Thus the infective stage was probably present in Ningi Creek some time between 18 and 25 January. This was precisely the time of heavy rain (fig. 2.1; rainfall), supporting a belief of many oyster farmers that the arrival of the disease is associated with "the first fresh".

The short time the infective stages were present probably reflects the fact that much of the creek water is changed with every tide. The movement of the buoy indicated that water over the commercial oyster trays mixed with Pummicestone Passage water at low tide and probably much of it also flowed out into Moreton Bay. There is a net northward movement of water through the passage (Co-ordinator General's Department 1982), so on an incoming tide not only has the creek water been heavily diluted but water that has remained in the passage is more likely to go north than return into the creek. In summary the creek is well flushed every day, and this probably accounts for the short time that the infective stages of the parasite were present.

The second infection in Ningi Creek, probably some time in April, was not associated with an equivalent period of heavy rain, though rain certainly occurred. The stage of the tide at the time of the rain may be important as this would affect the degree of change in salinity.

Late summer infections survived through the winter to continue the cycle the following spring. One might suspect that the number of overwintering infections would be related to the timing and severity of outbreaks the following summer. Predictions of outbreaks using this relationship would be invaluable to growers. However, recent results suggest there is little correlation. This, and the lack of infectivity of the spores to oysters, argue in favour of an intermediate host, the discovery of which would facilitate research on this interesting and economically important parasite.

Acknowledgments

The cooperation of oyster growers Dr C. Lucas, Mr N. Henry, and Mr A. Chard is much appreciated. Mr C. Boel and Mrs M. Barrett gave excellent field and histological assistance. The temperature and rainfall data were courtesy of the Australian Bureau of Meteorology. Dr L.R.G. Cannon made many valuable suggestions on an earlier draft of the manuscript. The project was supported financially by a grant from the Australian Fishing Industry Research Trust Account.

References

Co-ordinator General's Department. Queensland Government. 1982. Pummicestone Passage water quality and land use study. Brisbane: Co-ordinator General's Department, Queensland Government.

Perkins, F.O. and Wolf, P.H. 1976. Fine structure of *Marteilia sydneyi* n.sp. – haplosporidian pathogen of Australian oysters. *Journal of Parasitology* **62**: 528-38.

Wolf, P.H. 1972. Occurrence of a haplosporidan in Sydney rock oysters (*Crassostrea commercialis*) from Moreton Bay, Queensland, Australia. *Journal of Invertebrate Pathology* **19**:416-17.

Wolf, P.H. 1979. Life cycle and ecology of *Marteilia sydneyi* in the Australian oyster, *Crassostrea commercialis*. *Marine Fisheries Review* **41**:70-72.

3 *Pneumonema tiliquae* (Nematoda:Rhabdiasidae): A Reappraisal

R.J. Ballantyne

Introduction

Pneumonema tiliquae was described and figured in detail by Breinl (1913) from *Tiliqua scincoides* from north Queensland. He was not able, with the literature at his disposal, to place the worm in a genus, but considered it probably represented a new genus of the family Gnathostomidae. Johnston (1916) examined specimens from *T. scincoides* from the Sydney district (NSW) and Brisbane (Qld). He agreed with Breinl's findings and named the parasite *Pneumonema tiliquae* and stated that he hoped to classify it at a later date. Yorke and Maplestone (1926) gave a brief description with original figures and placed it in the family Rictulariidae of the superfamily Spiruroidea. Yamaguti (1961) followed Yorke and Maplestone's classification. Ballantyne and Pearson (1963) in a brief report, stated that there was an alternation of generations in the life cycle and that the parasitic adult was hermaphroditic and had uninucleate pharyngeal glands and therefore transferred *Pneumonema* to the family Rhabdiasidae of the superfamily Rhabdiasoidea. Baker (1981) redescribed *Pneumonema tiliquae* from seven specimens found in a museum collection in France.

Materials

The current redescription is based on material from north Queensland (1 host), Stradbroke Island (1 host) Brisbane area (50 hosts), northern New South Wales (1 host) and life cycle stages from laboratory-raised lizards. Worms were examined alive, heat-fixed, formalin-fixed, alcohol-fixed, cleared in glycerine, stained with aceto-orcein (Hirschmann 1962), and as serial sections (paraffin wax, stained Mayer's Haemalum and eosin) of whole worms and worms *in situ*. The original material of Breinl (1913), Johnston (1916) and Yorke and Maplestone (1926) could not be traced.

Details

Host: *Tiliqua scincoides* (Shaw, 1790), (Scincidae), blue-tongued lizard
Location in host: lungs
Locality: Queensland and New South Wales, Australia

Specimens: material deposited in the South Australian Museum, Adelaide, South Australia
Parasitic males absent, parasitic adult in the form of a female.

Results

Description of parasitic adult

Anterior end bent slightly dorsally, body slender, 5–12 mm long; width at pharyngo-intestinal valve 90–110 μm, at vulva 120–250 μm, at anus 60–75 μm. Tail 190–350 μm long, bluntly tapering at extremity with cuticle characteristically inflated; post-anal lip small.

Cervical papillae large, lateral, at level of nerve ring. Nerve ring 150–175 μm from anterior end. Two pairs of lateral expansions anterior to cervical papillae, anterior pair solid and darkened; posterior pair solid and darkened at external margin in old specimens, hollow and transparent in young specimens.

Thirty-six to 45 spines in 2 longitudinal rows on each lateral surface from cervical papillae to anterior end of uterus; remaining lateral surface smooth. Dorsal and ventral cuticle with fine longitudinal striations. Spines solid and darkened, alternating in position between each row and occurring in characteristic regions of large and small spines on each lateral surface. Anterior region of large spines all directed ventrally except for one large spine near the posterior end of this group, directed dorsally; middle region of 3 groups each with small spines followed by large spines, and a posterior region of all small spines, directed dorsally and ventrally in latero-dorsal and latero-ventral rows.

Small cuticle-lined ducts open on each lateral surface in 2 longitudinal rows from anterior to posterior end; in large spine region ducts open anterior and posterior to each spine.

Mouth opening small, in live worms the only sign of lip areas was the swelling or lobes associated with the papillae. In fixed material, mouth surrounded by 2 lateral lips. Each lip formed into 3 lobes on its internal margin and bearing 3 labial and 2 cephalic papillae as follows: paired labial and cephalic with cephalic externolateral to labial on subdorsal and subventral lobes, and a single labial internal and ventral to amphidial pore on lateral lobe. Amphidial glands not differentiated.

Buccal capsule globular 15 μm wide by 10 μm long, often collapsed in fixed specimens. Entrance to pharynx funnel-shaped and prominent. Cuticle of buccal capsule and pharyngeal entrance slightly thickened and darkened.

Pharynx clavate, 440–680 μm long; muscle fibres at right angles to lumen, those of anterior third very prominent. Pharyngeal glands uninucleate; nuclei 10 μm in diameter and 40–50 μm from posterior end of pharynx; subventral nuclei prominent, dorsal often indistinct, position of nuclei often obscured by torsion of posterior end of pharynx. Pharyngeal glands may appear very granular; dorsal gland opens into buccal cavity, subventrals open anterior to nerve ring.

Pharyngo-intestinal valve a cuticle-lined extension of pharynx 15–20 μm long; containing 5–7 nuclei; either projecting into lumen of intestine or contracted into base of pharynx in fixed material.

Intestine prominent, containing red blood cells or their products. Recto-intestinal valve poorly developed, containing 2 small lateral nuclei that were difficult to distinguish from intestinal nuclei and were often obscured by the rectal cells. Rectum a large cuticle-lined chamber 80–120 μm long, surrounded anteriorly by one dorsal and 2 latero-ventral binucleate rectal cells, and dorsally by a primordium of 5 nuclei (accessory primordium).

Phasmids small, lateral in midtail region. Excretory pore posterior to nerve ring, paired renette cells usually not prominent, lateral excretory canals absent.

Two complete female reproductive systems present. Vulva midbody, uteri opposed and ovaries reflexed. Anterior uterus, and non-reflexed area of ovary, shorter than similar posterior regions as forward expansion restricted by large body muscles. Oviduct folded on itself in midregion, forming a sperm reservoir. Ovary long, consisting of a distal germinal zone 300–350 μm long followed by a short synapse zone where sperm or ova production is alternately determined, and a long growth zone. Uteri contained a total of 200 or more eggs in fully mature adults. Eggs thin shelled, 85—100 μm by 45–65 μm, retained *in utero* until they contained fully formed first-stage larvae.

In section: cuticle and hypodermis 1 μm thick; hypodermal nuclei absent in dorsal cord; nuclei small in ventral cord; lateral cords well developed, in anterior body broad and thick between large muscles, remainder broader and flat; muscles well developed in anterior third of body, width 22–30 μm decreasing to 6 μm in posterior body; intestine thick walled, differentiated into several regions by variations in thickness (10–30 μm) and nuclear number (10–35) per cross section, internal margin of some cells extended into lumen of intestine forming longitudinal ridges giving the internal wall a folded appearance in cross section.

Life cycle

A rhabditoid free-living generation developed through 4 moults in the host's faeces to produce adult males and females in 24–36 h. Adult females did not have a fully developed vulva and eggs hatched *in utero* and ate out the female body. Infective third-stage larvae (500–900 μm long) were ensheathed and were found in the faeces 84–96 h after defaecation.

Laboratory-raised lizards were infected orally and third-stage larvae were found in the body cavity 24–48 h after infection. Larvae developed and underwent two moults in the body cavity. The third- and fourth-stage

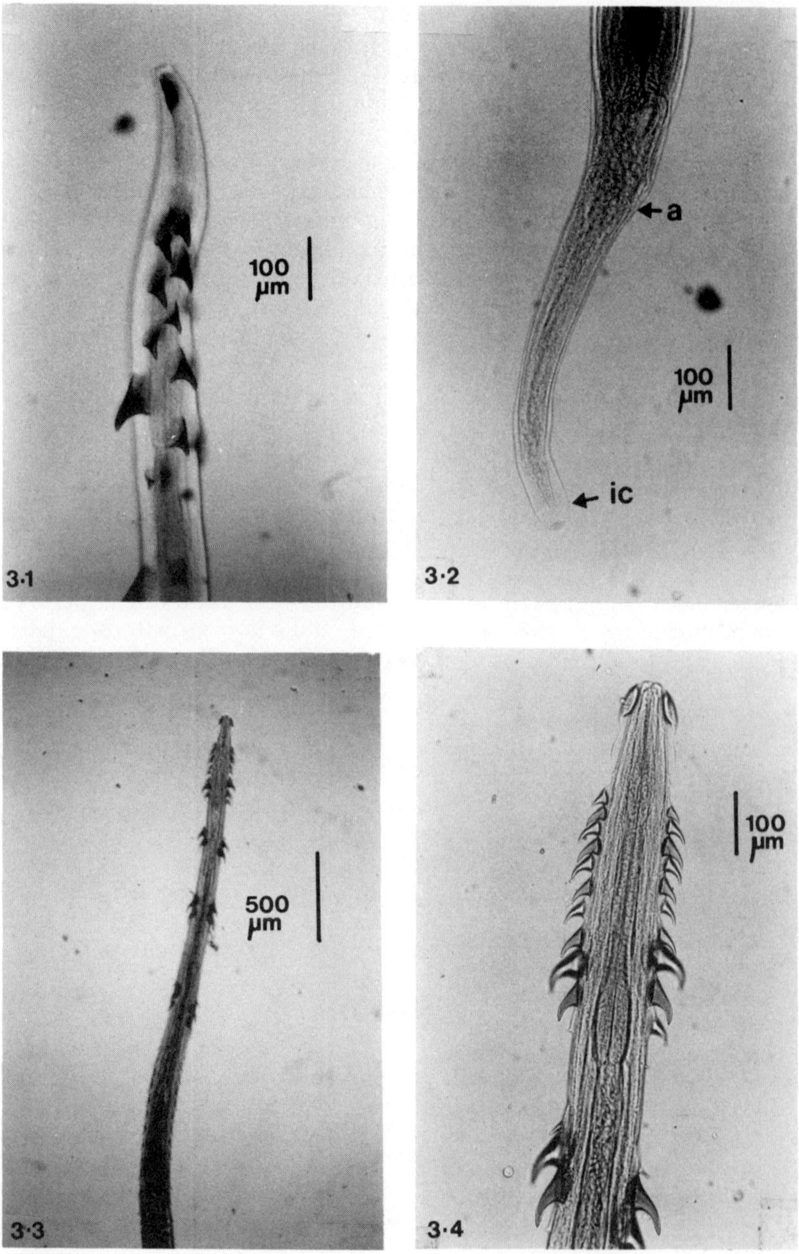

Figures 3.1–3.4 *Pneumonema tiliquae* mature adult. **3.1** Anterior lateral — bent dorsally, spines alternate in position in each row, one dorsally directed spine. **3.2** Posterior lateral — cuticle characteristically inflated (*ic*), small post-anal lip (*a* = anus). **3.3** Anterior ventral — showing complete spine pattern. **3.4** Higher power of 3.3, anterior lateral expansion hardened, cephalic papillae anterior to first spine.

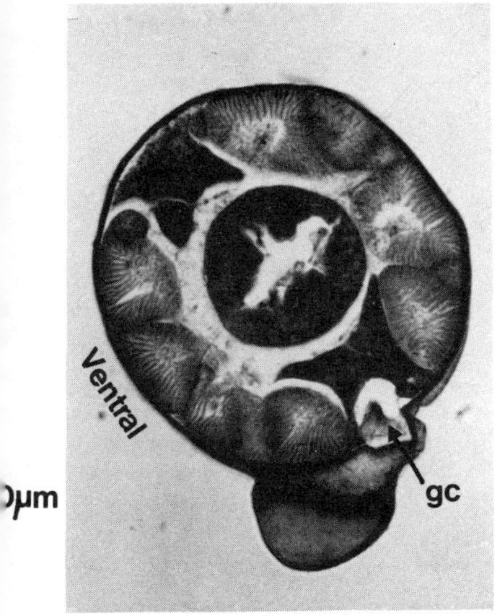

3.5

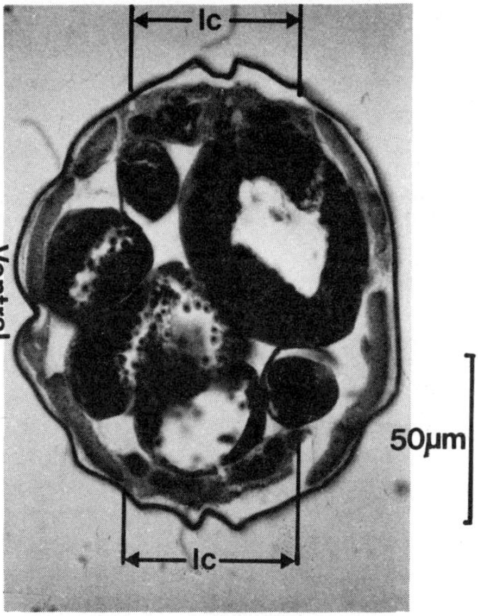

3.6

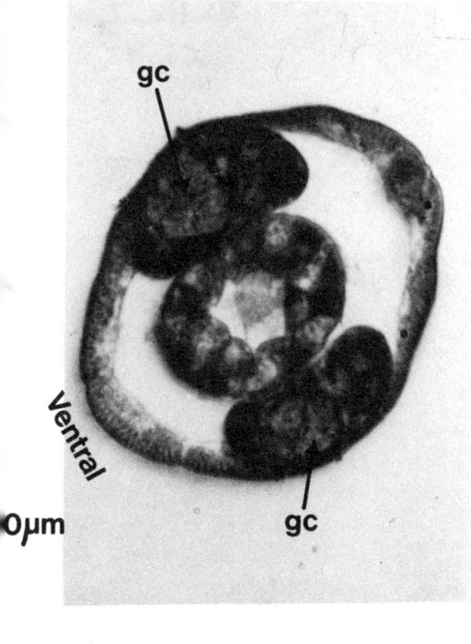

3.7

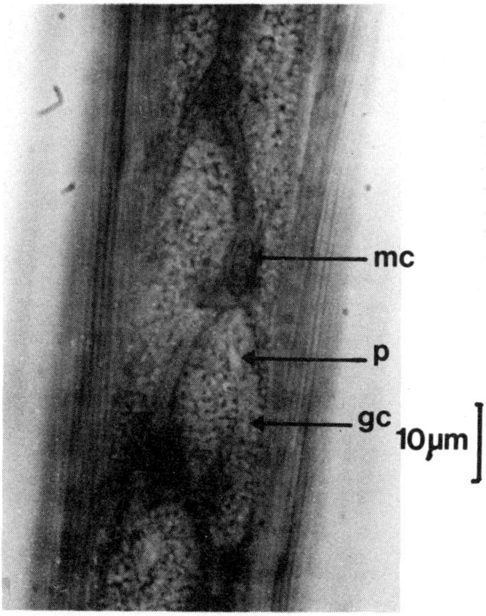

3.8

Figures 3.5–3.8 *Pneumonema tiliquae*. **3.5** Transverse section, adult. Posterior to pharyngo-intestinal valve at level of first large spine — large muscles, well developed lateral cord, gland cell (*gc*) empty. **3.6** Transverse section, adult. Level of posterior oviduct — muscles small, lateral cord (*lc*) flattened. **3.7** Transverse section, immature adult from body cavity, level similar to 3.5. **3.8** Lateral cord, surface view, in region posterior to pharyngo-intestinal valve, prior to spine formation — large gland cells (*gc*), pore (*p*), midcord nuclei (*mc*), longitudinal striations on cuticle.

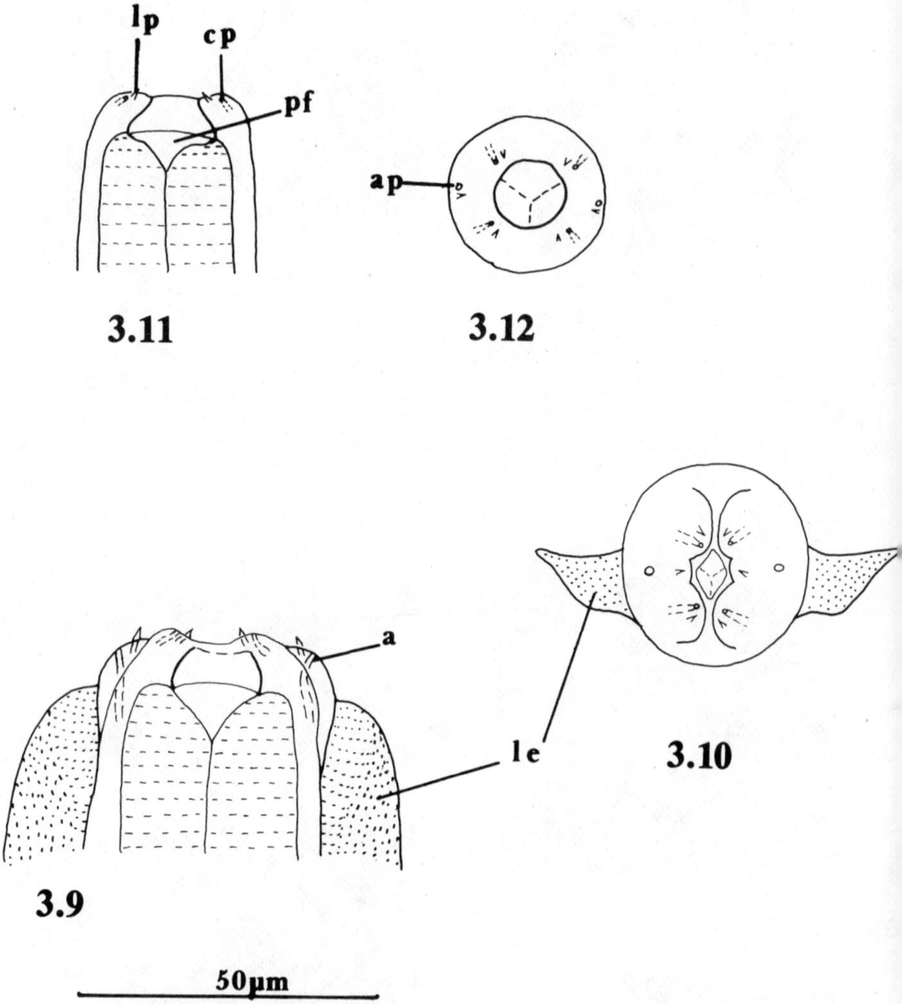

Figures 3.9–3.12 *Pneumonema tiliquae*, anterior end. *a* = amphidial duct; *ap* = amphidial pore; *lp* = labial papilla; *cp* = cephalic papilla; *le* = lateral expansion; *pf* = phargyngeal funnel. **3.9** Ventral, young adult 3.5 mm long. **3.10** *En face*, adult 8 mm long, buccal capsule collapsed, lips formed. **3.11** Immature adult from body cavity, lateral. **3.12** Immature adult from body cavity, *en face*.

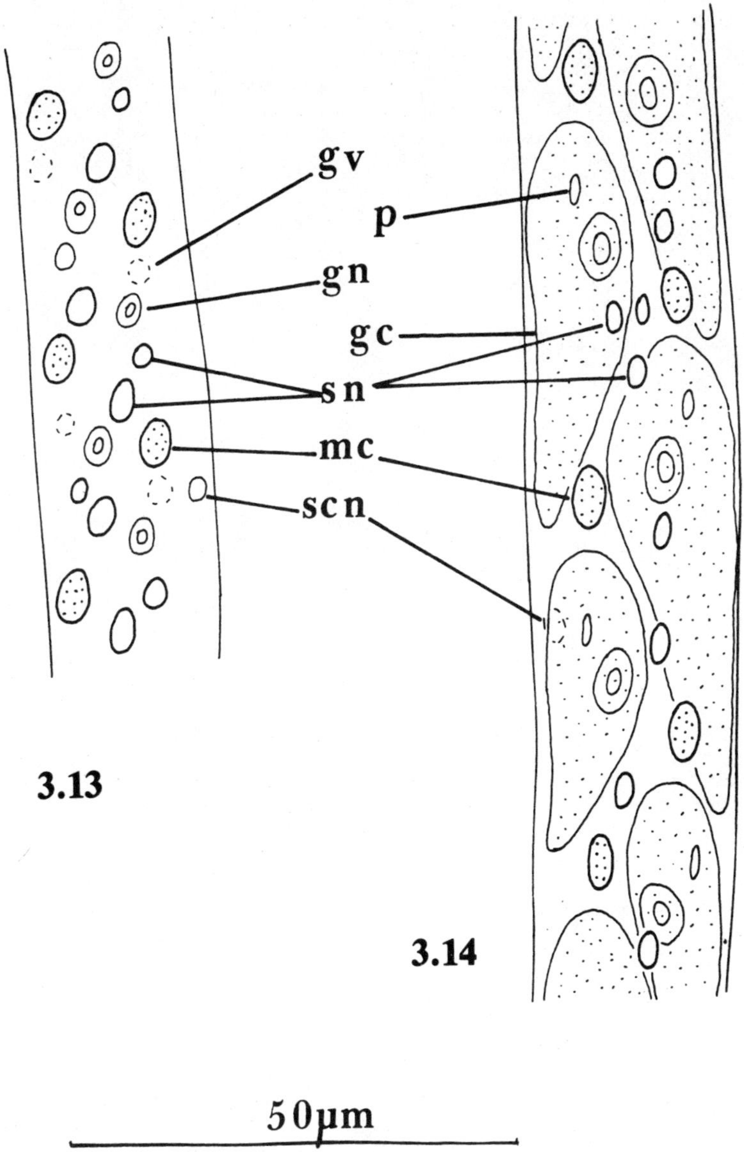

Figures 3.13–3.14 *Pneumonema tiliquae* — lateral cord of parasitic generation, posterior to pharyngo-intestinal valve. *gv* = gland vacuole; *gn* = gland nucleus; *gc* = gland cell; *mc* = midcord nuclei; *p* = pore; *scn* = scattered nuclei; *sn* = surface nuclei. **3.13** Mid-fourth stage. **3.14** Immature adult prior to spine formation.

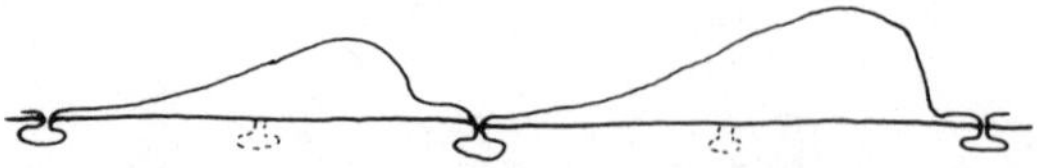

3.15

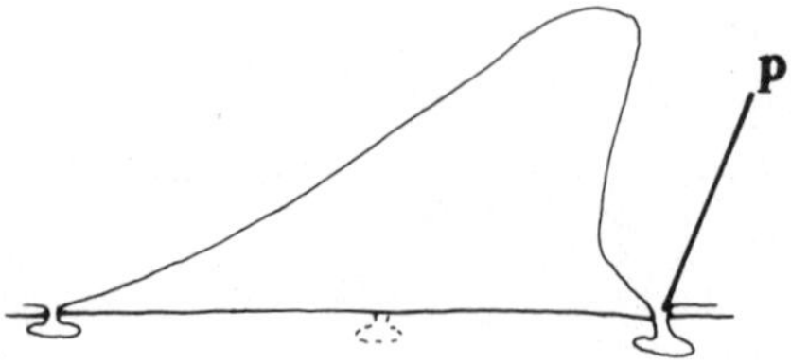

3.16

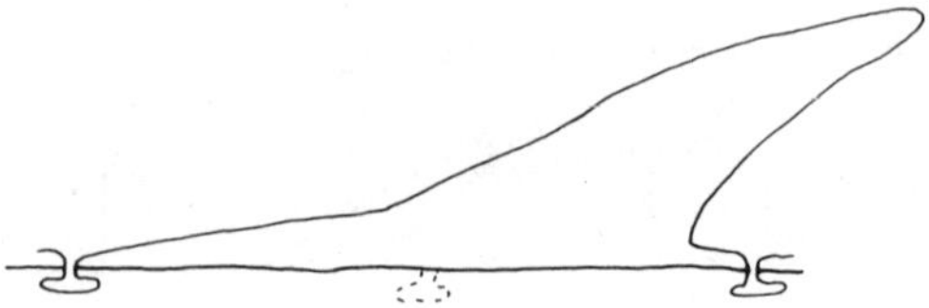

3.17

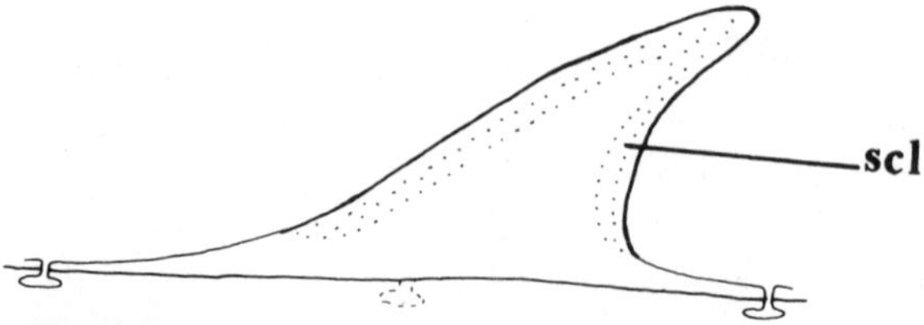

3.18

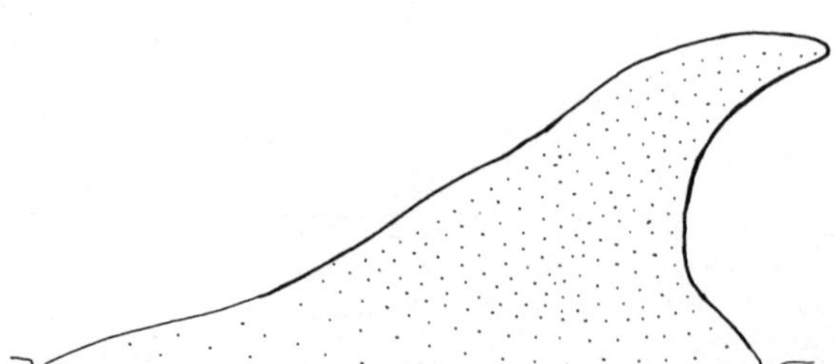

3.19

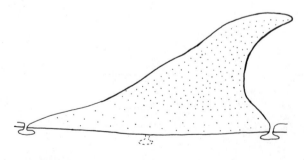

3.20

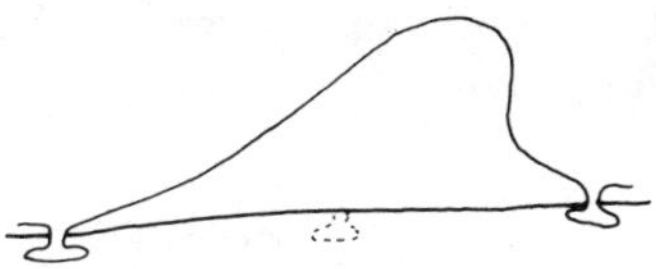

3.21

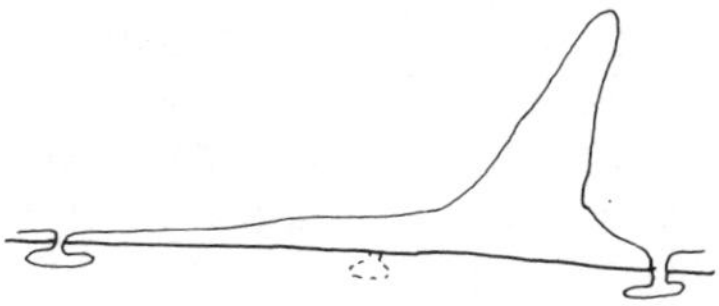

3.22

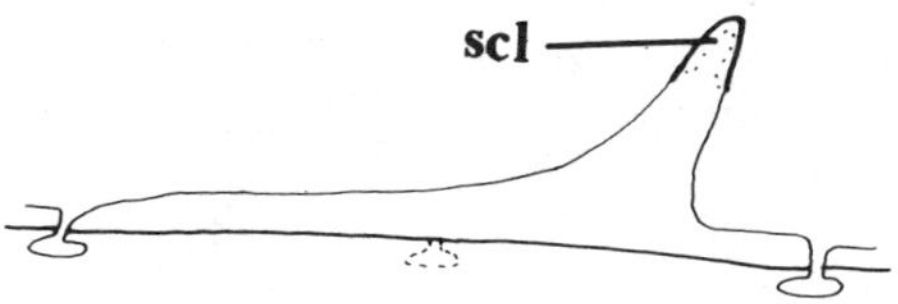

3.23

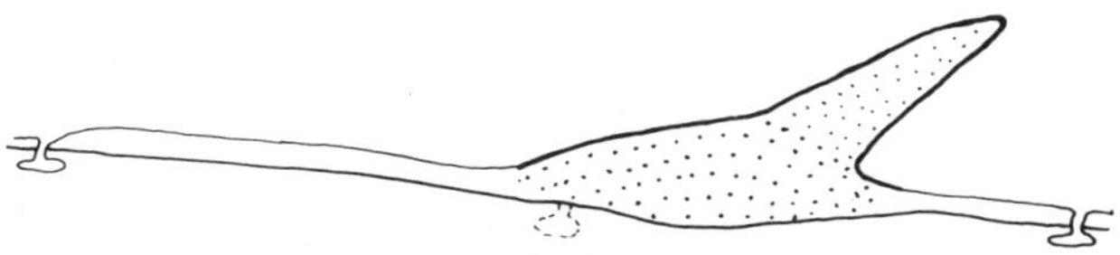

50μm

3.24

Figures 3.15–3.24 *Pneumonema tiliquae* spine formation in parasitic adult. *p* = pore; *scl* = sclerotised thickening. **3.15–3.19** Large spines. **3.20–3.24** Small spines. **3.15, 3.20** Initial dome formation, worm 3.15 mm long. **3.16,3.21** Cuticle forming into spine shape, worm 3.3 mm long. **3.17,3.22** Cuticle formed into spine shape, worm 3.9 mm long. **3.18,3.23** Spines beginning to fill and harden, worm 4.35 mm long. **3.19,3.24** Spines fully formed in mature adults, worm 7.2 mm long.

sheaths were retained for a short period and immature adults were found 7–8 days after infection.

Subsequent development depended on whether adult worms were present in the lungs. If adult worms were present in the lungs, development ceased when the germinal zone of the reproductive system contained 28–30 germinal nuclei and the immature adults remained in the body cavity. If adults were not in the lungs, division continued in the germinal zone of the reproductive systems producing sperm and the worms entered the lungs 12–14 days after infection.

Description of immature adults in the body cavity

Body slender 2.3–2.8 mm long; width at pharyngo-intestinal valve 60–65 μm, at vulva 55 μm, at anus 35–40 μm. Tail 115–150 μm long, tapering to a point, cuticle not inflated; post-anal lip small.

Body cuticle not inflated, with fine longitudinal striations and a small longitudinal ala (ridge) on each lateral surface. Lateral cords well developed with small cuticle-lined ducts opening on each lateral surface (ducts appeared during fourth moult).

Six labial and 4 cephalic papillae on rim of buccal cpasule. Amphidial ducts prominent, but glands not differentiated. Cervical papillae, buccal capsule (not collapsed), pharynx, pharyngo-intestinal valve, recto-intestinal valve, rectum, rectal cells, accessory primordium and phasmids as described for mature adult.

Renette cells large and prominent, excretory pore slightly posterior to level of nerve ring. Body muscles evenly developed throughout the length of the body. Intestine multinucleate, cell membranes clearly defined, granular in appearance, not obviously differentiated into regions by variation in nuclear number per cross section. Pseudo-coelomocytes large (5 μm in diameter) and clearly differentiated, 4 ventrally and evenly spaced between the anterior reflexion of the ovary and the pharynx, and 1 dorsally posterior to the reflexions of the posterior ovary.

Two complete reproductive systems present. Vulva mid-body; uteri opposed and may contain effete cells; oviduct usually folded on itself in midregion; ovary long and reflexed, distal end containing 20–30 germinal cells.

Development in the lungs

(figs. 3.15–3.24)

The exact method of entry into the lungs was not determined. Immature adults in the body cavity did not feed on blood, whereas all immature adults taken from inside the lungs had fed on blood. Two immature adults removed from the outside wall of a lung contained blood suggesting that entry may have been across a blood vessel during or after feeding.

Growth continued in the lungs and nuclear division occurred only in the intestine and in the germinal cells of the ovary.

The spines and lateral expansions soon appeared and were fully formed in 2 days when the body length was 4–5 mm and ova were ready for fertilization. The first indication of spines was the appearance of almost transparent hemispherical domes in the cuticle between the pores of the lateral cord. These domes corresponded to the position and size of spines in the adult. Gradually the apex of each dome formed into a posteriorly directed point. The foregoing development produced a rose-thorn-like spine with an oval base. The large spines began to fill and darken on the inner surface of the cuticle and were solid structures when the worm was 3.5–4.0 mm long. The smaller spines began to fill and darken from the apex and in adult worms were solid structures with a pronounced basal area.

The number of large spines was constant, whereas variation in number occurred in the small spines in the mid and posterior region.

The lateral expansions at the anterior end and the characteristically inflated cuticle of the tail formed at the same time as the spines. The lateral expansions formed in a similar manner to the spines; the anterior pair became solid and darkened whereas the posterior pair remained hollow except for some filling at the outer margin in older worms.

The anterior body muscles developed at the same time as the filling and darkening of the spines. The muscle cells did not divide but enlarged in proportion to the size of the spines. In the midregion of small and large spines the muscles were uniformly large although not as large as the muscles associated with the anterior region of large spines. In the posterior region of spines where all the spines were small the muscles became gradually thinner towards the posterior end of the region. All muscles increased only slightly in thickness during subsequent adult growth.

As the immature adults grew all parts of the anterior extremity, except the buccal capsule, increased in width. This growth resulted in an increase in the tissue surrounding the buccal capsule. In fixed material the anterior extremity contracted unevenly owing to the uneven rigidity produced by the lateral expansions, resulting in the formation of two lateral lip areas. In live worms the only sign of lip areas was the swellings associated with the papillae.

Egg-producing worms continued to grow and by days 17–20 were 6.5–8 mm long. Growth occurred mainly in the posterior body as shown by the position of the vulva changing from 57% to 44% of the body length as worms increased in size (distance measured from the anterior end). The spines and lateral expansions apparently restricted the growth of the anterior region.

Development of the lateral cords

During the early fourth stage, and without any increase in the size of the cord, the nuclei differentiated into the final pattern. All nuclei, except those deep in the cord, which belonged to the future gland cells, stained intensely.

The nuclei were characteristically arranged in recurring groups of four nuclei with a vacuole anterior to the gland nucleus in each group. Each group consisted of 1 gland nucleus, 1 midcord nucleus and 2 surface nuclei.

During the fourth stage the cord widened to 30 μm with a slight increase in nuclear size. The gland vacuole developed into a duct to the surface and became cuticle-lined at the fourth moult. Some of the surface nuclei divided at the fourth moult.

While the immature adult remained ensheathed the gland cells became prominent especially in the anterior half of the body. As the immature adult grew the surface nuclei lost their pattern and the scattered dorsal and ventral nuclei became apparent as cells rather than nuclei.

In the adult, the shape of the cord was altered by the development of the anterior body muscles and the reproductive system. Some gland cells near spines appeared to be empty whereas those of the posterior body were flattened by the reproductive system and it was difficult to ascertain their contents.

Discussion

The current study using life cycle studies in association with adult taxonomy has clarified several points. The variation in body length given in previous descriptions occurs because adult worms continue to grow for some time after egg production begins. Likewise ova formation and egg development are integrated with the sperm cycle in the gonad and eggs are retained *in utero* until they contain fully developed larvae.

The lateral lips are a constant artifact, resulting from uneven contraction of the anterior end during fixation. The buccal capsule often collapses during fixation as the muscles of the pharynx contract and draw the buccal capsule tissue into the triradiate pharyngeal entrance. The contracted condition is easily recognized in fixed material, as the muscle fibres of the anterior end of the pharynx are not at right angles to the lumen but are angled inwardly and posteriad. Constant artifacts are not new in histology as it is well known that different fixatives penetrate at different rates into different tissues (Culling 1974) and the problem of artifacts is so widespread that Thompson and Luna (1978) published a text illustrating artifacts encountered in the preparation of microscopic tissue sections.

The use of aceto-orcein as a stain not only aids in the detailed study of cells but aids in the study of papillae. The cephalic papillae stain intensely and appear morphologically different to the labial papillae. The current description is the first to record the presence of 10 papillae; Breinl (1913) recorded 4 whereas Baker (1981) recorded 6.

The term lateral expansion was used instead of lateral alae to call attention to their development which occurred at the same time and in the same manner as the spines. The thickening of the anterior expansion gave it a folded appearance.

One of the most interesting features of *P. tiliquae* is the presence of body spines. Body spines have been recorded from some marine nematodes and the parasitic order Spiruroidea (Hyman 1951). Oswald (1958) found that the spines of *Rictularia coloradensis* formed at the last moult. The spines of *P. tiliquae* are formed during adult growth, after the last moult and after entry of the immature adults into the lungs. The anterior body spines of *P. tiliquae* although large and directed posteriorly do not appear to aid in attachment to the alveolar walls. *In situ* there was no evidence either of contact between the spines and the alveolar tissue or of host reaction to the presence of worms.

The rhabdiasids are unusual parasitic nematodes as they have an obligatory free-living generation in their life cycle. *P. tiliquae*, as well as maintaining the free-living generation, retained immature adults in the body cavity if adults were in the lungs. This appears to be a major development as the body cavity is a vastly different habitat to the lungs and must require a different physiological way of life. Although the free-living and parasitic generations appear morphologically distinct, closer examination shows they contain the same basic cell pattern for each system. The lateral cords show similar development until the early fourth stage, when development stops in the free-living generation, but continues in the parasitic generation to form large cells with cuticle-lined ducts to the surface.

In rhabdiasids there appears to be an evolutionary relationship between inflated cuticle, enlarged body muscles and lateral cord development. During the current study numerous rhabdiasids besides *P. tiliquae* were examined from lizards, snakes and frogs. The cuticle was either inflated as in *Rhabdias* spp. (Baker 1978) or loosely applied to the body as in *Entomelas* spp. The anterior body muscles were enlarged in *Rhadias* spp. from lizards, and poorly developed in all other species of *Rhabdias* examined. The lateral cords had the same basic structure as in *P. tiliquae* including pores to the surface but the cells were similar to those of the posterior body region where the cells although large, appeared to have lost their secretory ability. The cuticle-lined ducts were present in all species examined and were best seen in live and freshly fixed material. All previous studies on rhabdiasids, except Singh and Ratnamala (1977) have overlooked these ducts. Singh and Ratnamala described pore-like openings in the cuticle of a rhabdiasid, and on the basis of these pores and the size of the buccal capsules established the genus *Shorttia*. Baker (1980) synonymized the genus with *Rhabdias*. The only other record of cuticle-lined pores in the cuticle of animal parasitic nematodes known to the author, is that of Bhaibulaya (1968), who described a pair of phasmid-like pores anterior to the anus in *Angiostrongylus mackerrasae*. Many marine nematodes have been described as having large gland cells in the lateral cords with ducts opening through the cuticle (Hyman 1951). *P. tiliquae* with its precisely controlled cuticle inflation followed by spine and muscle development coupled with its restriction to a single host species appears to represent a stable gene pool. *Rhabdias* with its uncontrolled cuticle inflation, variable muscle development plus cosmopolitan distribution and wide range of hosts appears to represent an unstable gene pool. The rhabdiasid genera appear to be quite different to *Strongyloides* and *Parastrongyloides* and may not even have a common ancestry.

Acknowledgments

The assistance of the following individuals in the collection of hosts or specimens is gratefully acknowledged: J. Ballantyne, A.C. Chabaud, J. Hickman, M.W. Lankester, J.C. Pearson, J.F.A. Sprent who provided the material that initiated my interest in rhabdiasids, and the late M.J. Mackerras. I am greatly indebted to Dr John Pearson for the initial suggestion to culture the first-stage larvae and for his continued assistance during my studies at the University of Queensland. This study was supported by the University of Queensland and a Commonwealth Postgraduate Award from 1968–1970.

References

Baker, M.R. 1978. Morphology and taxonomy of *Rhabdias* spp. (Nematoda: Rhabdiasidae) from reptiles and amphibians of southern Ontario. *Canadian Journal of Zoology* **56**:2127–41.

Baker, M.R. 1980. Revision of *Entomelas* Travassos, 1930 (Nematoda:Rhabdiasidae) with a review of genera in the family. *Systematic Parasitology* **1**:83–90.

Baker, M.R. 1981. Redescription of *Pneumonema tiliquae* Johnston, 1916 (Nematoda:Rhabdiasidae) from an Australian skink. *Proceedings of the Helminthological Society of Washington* **48**:159–62.

Ballantyne, R.J. and Pearson, J.C. 1963. The taxonomic position of the nematode *Pneumonema tiliquae. Australian Journal of Science* **25**:498.

Bhaibulaya, M. 1968. A new species of *Angiostrongylus* in an Australian rat, *Rattus fuscipes. Parasitology* **58**: 789–99.

Breinl, A. 1913. Nematodes observed in north Queensland *Report of the Australian Institute of Tropical Medicine* (1911):39–46.

Culling, C.F.A. 1974. *Handbook of histopathological and histochemical techniques*. 3rd ed. London: Butterworths.

Hirschmann, H. 1962. The life cycle of *Ditylenchus triformis* (Nematoda:Tylenchida) with emphasis on post-embryonic development. *Proceedings of the Helminthological Society of Washington* **22**:115–23.

Hyman, L.H. 1951. *The invertebrates; Acanthocephala, Aschelminthes, and Entoprocta. The pseudocoelomate Bilateria*. New York: McGraw-Hill.

Johnston, T.H. 1916. A census of the endoparasites recorded as occurring in

Queensland, arranged under their hosts. *Proceedings of the Royal Society of Queensland* **28**:31–79.
Oswald, V.H. 1958. Studies on *Rictularia coloradensis* Hall, 1916 (Nematoda:Thelaziidae). II. Development in the definitive host. *Transactions of the American Microscopical Society* **77**:413–22.
Singh, S.N. and Ratnamala, R. 1975 (issued 1977). On a new genus and new species of rhabdiasoid nematode *Shorttia shorttia* n.g., n.sp. infesting lungs of amphibians. *Indian Journal of Helminthology* **27**: 132–38.
Thompson, S.W. and Luna, L.E. 1978. *An atlas of artifacts encountered in the preparation of microscopic tissue sections*. Springfield, Illinois: Charles C. Thomas.
Yamaguti, S. 1961. *Systema Helminthum*. III. The nematodes of vertebrates. Parts I and II. New York: Interscience Publishers.
York, W. and Maplestone, P.A. 1926. *The nematode parasites of vertebrates*. London: Churchill (Hafner Publishing Co. New York, 1962).

4 The Paranephridial System in the Digenea: Occurrence and Possible Phylogenetic Significance

J.C. Pearson

The digenean excretory system figures prominently in the major classification of La Rue (1957), and in the major, if informal, revision of La Rue's classification by Cable (1974). In his ICOPA V address in 1982, Cable reiterated his view that the relationship between the excretory system and the cercarial tail, rather than the form of the excretory bladder, provides a truly phylogenetic basis for primary division of the Digenea into major taxa to replace the orders of La Rue's (1957) system. He went on to say that all aspects of digenean biology could contribute to refining that system. One such aspect, largely ignored, is the paranephridial system. It is the aim of this account to point out how widespread a paranephridial system is, and to suggest that its occurrence in four of the six lines of descent in Cable's (1974) scheme bespeaks an early origin in the Digenea, with the corollary that absence in advanced groups may indicate loss.

Before advancing arguments in support of this thesis, it is necessary to define the paranephridial system and describe some of its forms and variations. If the protonephridial system proper "consists of a tube or tubules opening distally via a nephridiopore at the surface of the animal and ending proximally in one or, more usually, many terminal structures" (Wilson & Webster 1974) bearing a tuft of cilia and a cap nucleus, hence flame cells (fig.4.1), then the paranephridial system may be defined as a series of outgrowths from the bladder, bladder arms, and primary collecting tubules, that may ramify and anastomose, but do not possess flame cells (fig.4.1). The system may begin to develop in the cercaria or in the metacercaria, but is best developed in the adult.

The paranephridial system takes a variety of forms, from those with few and simple outgrowths, as in the plagiorchioid *Cephalogonimus* (fig.4.2), to those with numerous vessels that ramify and anastomose throughout the body, as in clinostomes (fig.4.3) or that enlarge to form vast spaces, veritable body cavities, as in strigeids (fig.4.4).

Additionally, the system may be present in one member but absent from another member of a group. At the generic level this is seen, for example, in philophthalmids, in which it is well developed in *Parorchis* but absent from *Philophthalmus*. At the specific level, it is present in *Echinochasmus perfoliatus* but not in *Echinochasmus japonicus*, according to Komiya and Tajimi (1941).

Diplodiscid and paramphistomid flukes were included by Braun (1893)

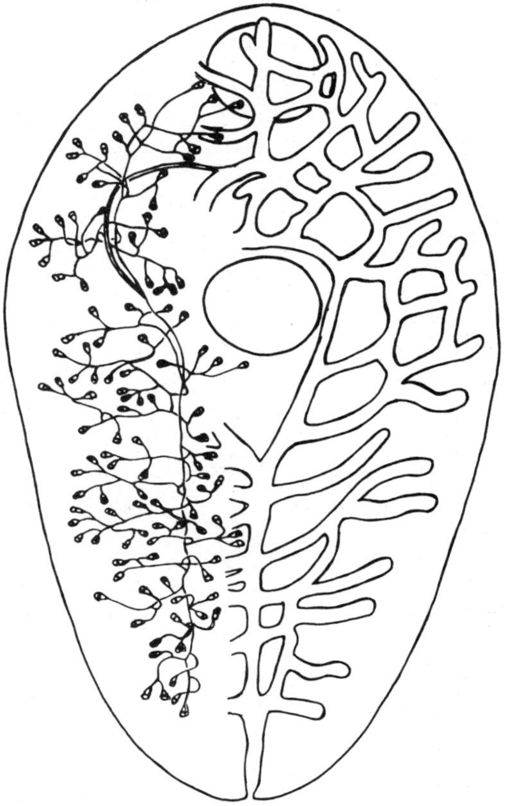

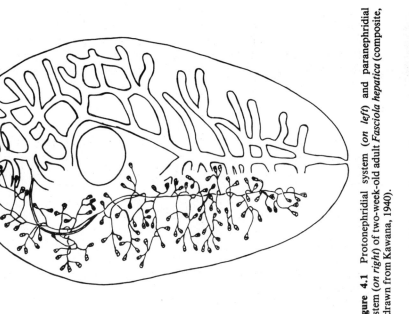

Figure 4.1 Protonephridial system (*on left*) and paranephridial system (*on right*) of two-week-old adult *Fasciola hepatica* (composite, redrawn from Kawana, 1940).

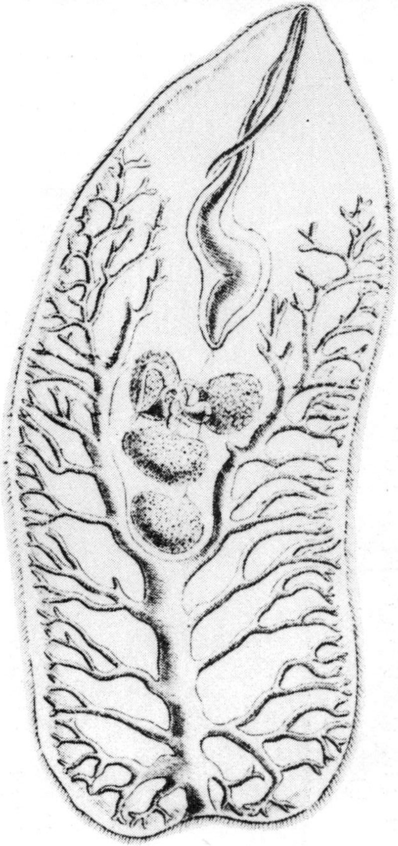

Figure 4.2 Poirier's figure of *Cephalogonimus* (from Braun 1893).

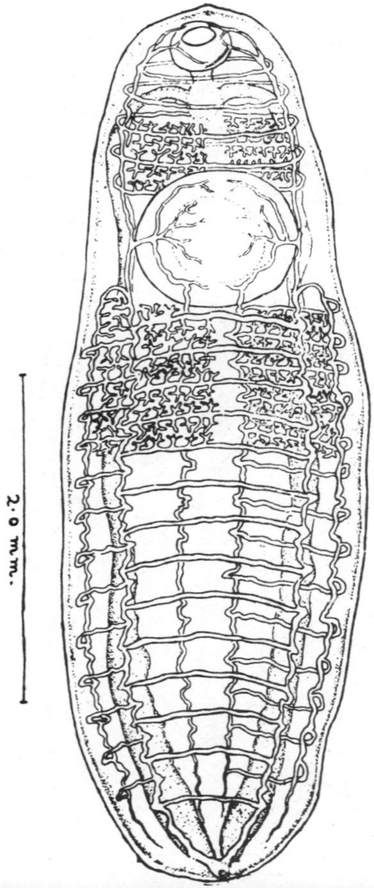

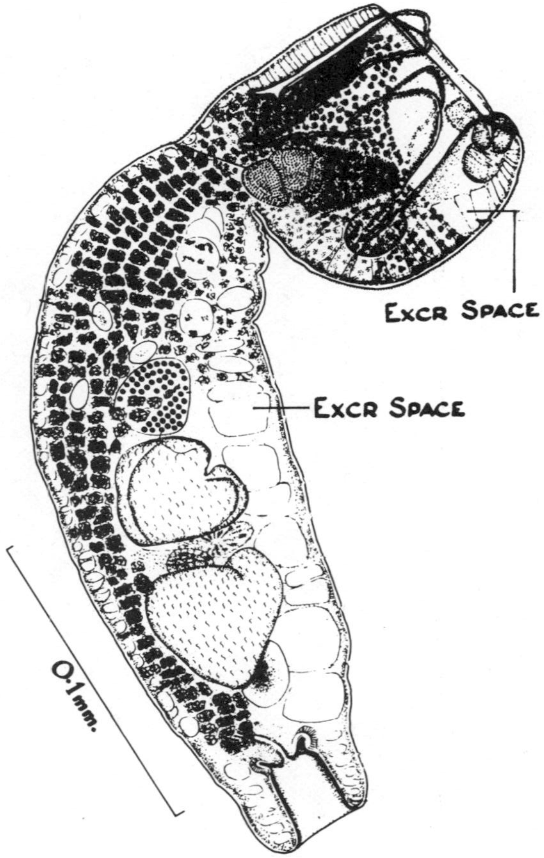

Figure 4.4 *Strigea* (author's drawing). Excr. space = spaces

among forms showing modified excretory bladders, but Looss (1902) pointed out that the accessory vessels in paramphistomids are not connected to the excretory system. He differentiated between the separate blind plexus of paramphistomoids, which he termed a lymphatic system, and vessels connected to the excretory system proper, which he had earlier termed a reserve bladder (= reserve excretory system, now replaced by paranephridial system, derived from Reisinger and Graack's [1962] term, paranephridial plexus).

The lymphatic system of paramphistomoids is held to be distinct and separate from the excretory system (Cheng 1966) but Willey (1936) described the diplodiscid *Cercaria poconensis* with branched main excretory tubules, an observation confirmed by Hussey (1941) and by Kuntz (1952, p.68) who says "the diverticula increase in size and number, and apparently are carried through subsequent stages of development". Additionally, Fain's (1953) figure (plate 3, fig. 1) of the excretory system of the presumed diplodiscid *Cercaria nigrita* (fig. 4.5) is remarkably similar to the figure by Walter (in Braun 1893) of the adult diplodiscid *Diplodiscus subclavatus* (fig. 4.6). Finally Willey (1954) claimed that the lymphatic system of the paramphistomid *Zygocotyle lunata* arises from the excretory system, but loses all connexion with it in fully grown flukes. This claim has been neither confirmed nor confuted, but the weight of present evidence favours interpreting the accessory vessels of paramphistomoids as paranephridial.

If the vessels in paramphistomoids are paranephridial although blind, what of similarly blind vessels known in microscaphidiids, glyiauchenids, and lepocreadioids?

In microscaphidiids, Stunkard (1943) found, but did not figure, a paranephridial system but no separate lymphatic system in *Dictyangium*, whereas Ozaki (1937) figured both branching, anastomosing bladder arms and separate blind tubules in *Hexangium*,. The family has a paranephridial system, but whether the lymphatic system of some members originates from it must wait on studies of development.

In the developing cercaria of the lepocreadiid *Neopechona pyriforme*, Stunkard (1969) observed that from the fusion of the two protonephridia at the tail-body junction, a long, tubular outgrowth developed in the body. Thus, bladders of this form in species of *Opechona* and *Neopechona* (Kφie 1975) and of *Lepocreadium* (Bartoli 1967) may indicate vestiges of a paranephridial system. I did not find such an outgrowth from the bladder in *Tetracerasta* and *Stegodexamine*, nor did Watson (1984) who also examined them alive. Whether the lymphatic vessels described by Manter (1940) in *Opechona* and *Pseudolepidapedon* and by Manter (1937) in species of the homalometrid *Apocreadium*, and by Ozaki (1937) in gyliauchenids, and by Manter (1947) in megasolenids, two families placed by Cable and Hunninen (1942) close to lepocreadiids, can only be decided when the development of their excretory systems is known.

When information on occurrence (see Appendix) is superimposed on

Cable's (1974) scheme of six major lines of digenean descent (fig.4.7) it can be seen that of the 56 families included, 17 (marked + +) possess a well-developed paranephridial system in most or all members; 4 families (marked +) possess a well-developed system in at least some members and 13 families (marked + ?) possess what are taken to be vestiges. Thus, there are indications of a paranephridial system in 34 of 56 families. It follows that the paranephridial system is a major feature of the Digenea, and as such should figure prominently in general accounts of digenean morphology.

As to phylogenetic significance, the simplest way of explaining the distribution of the paranephridial system is to assume its origin in the common ancestor of at least lines *II, III, IV* and *V*, followed by its reduction in the schistosomatoid branch of line *IV*, and in most members of lines *II* and *III*, and branch 2 of live *V*. This leaves two lines, *I* and *VI*, represented by two small and primitive families, bivesiculids and transversotrematids, without indications of a paranephridial system.

The law of parsimony is here invoked not in an absolute way to swell the ranks of those groups with a paranephridial system, but rather as a sort of null hypothesis to be challenged and perhaps controverted in doubtful cases, such as lepocreadioids. As Professor Cable has pointed out to me, it may well be that different systems of vessels have been called into being by different challenges.

Absence of a paranephridial system in schistosomatids may be read as loss in view of the great development of that system in strigeoids and its presumed vestigial state in some spirorchiids. And this, together with the interpretation of the schistosome life cycle as a modified three-host life cycle from which the original definitive host has been lost (La Rue 1951; Pearson 1972) militate against acceptance of Richard's (1971) hypothesis that the schistosome cercaria is the most primitive type.

Within the family Philophthalmidae two genera, *Parorchis* and *Echinostephilla*, have a paranephridial system and an interrupted collar with spines. Both of these features are characteristic of echinostomes, the group closest to philophthalmids. Three genera, *Philophthalmus, Pygorchis*, and *Cloacitrema*, lack both paranephridial system and collar. It appears, then, that the paranephridial system has been lost in some members of the family. And as it is present in *Parorchis* and *Echinostephilla* but absent from *Pygorchis* and *Cloacitrema*, all four of which inhabit the cloaca, then its occurrence does not appear to be related to habitat within the host.

In branch *1* of line *V* a paranephridial system is well developed in both the largely freshwater Notocotylidae and the marine Rhabdiopoeidae and in branch *3* in the largely freshwater Paramphistomidae and in the marine Microscaphidiidae. It would appear that salinity does not bear on the expression of the system.

Among plagiorchioids the paranephridial system is most highly developed in renicolids and, by comparison, is reduced or absent in other

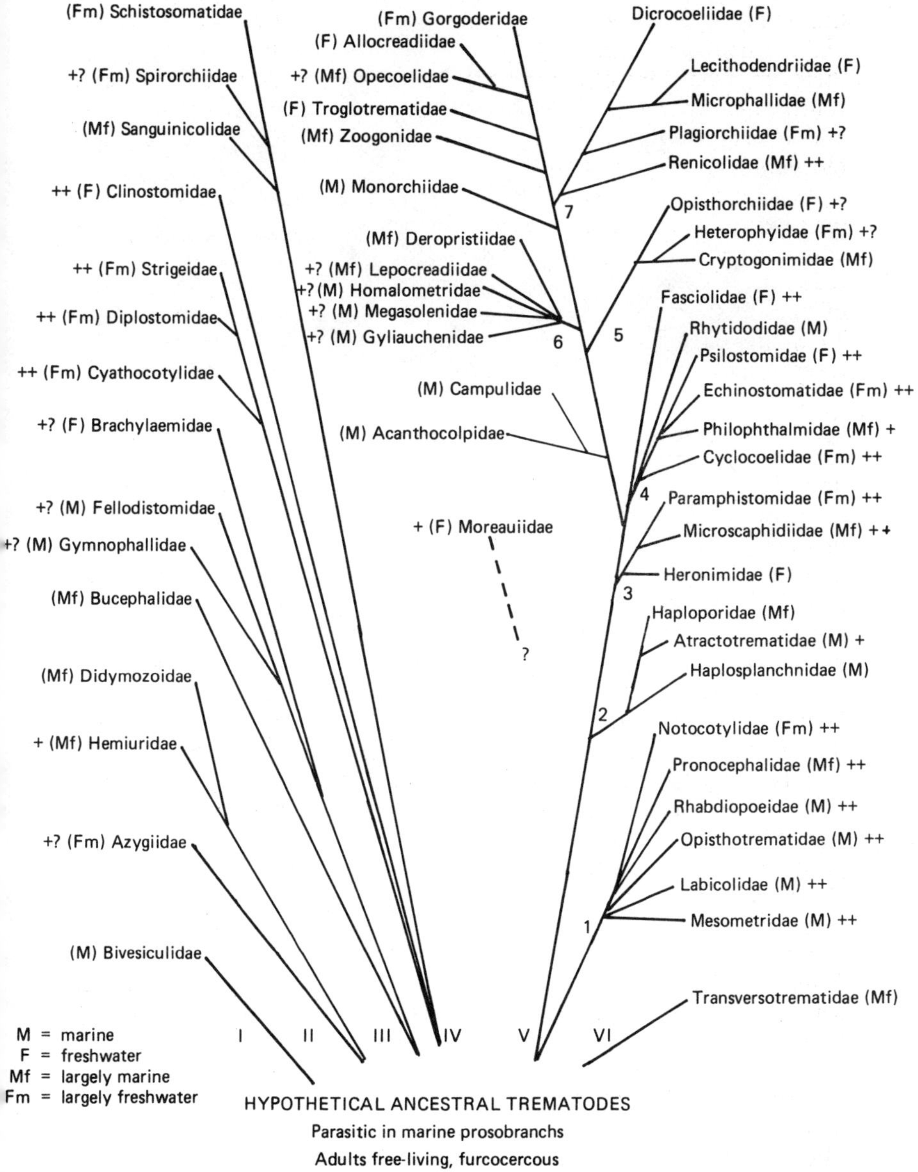

Figure 4.7 Occurrence of paranephridial system shown on a modified version of Cable's (1974) phylogeny of the Digenea. + + = paranephridial system well developed in most members; + = well developed in some members; + ? = with vestiges in some members.

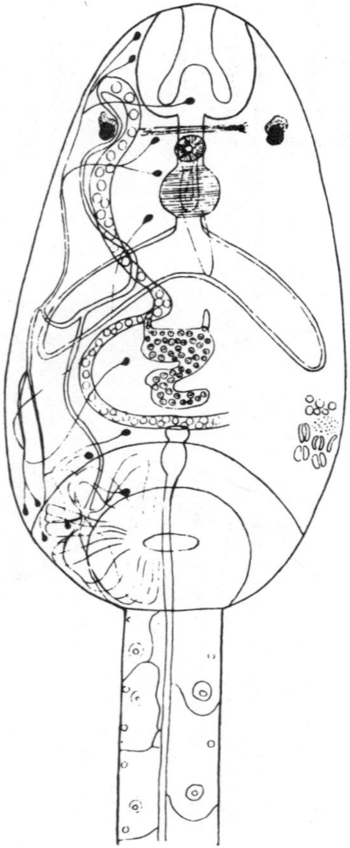

Figure 4.5 *Cercaria nigrita* (from Fain 1953).

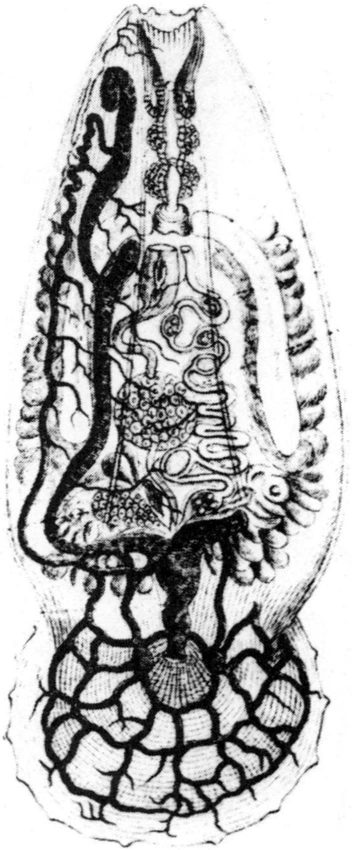

Figure 4.6 Walter's figure of *Diplodiscus subclavatus* (from Braun 1893).

families. This difference, together with variations in cercarial features, such as body size, degree of development of paranephridial system, and presence or absence of corpuscles in the excretory system, stylet, and frank penetration glands, led Pearson (1972, p.178) to make the surprising suggestion that among plagiorchioids renicolids may be the primitive group in which the stylet makes its first appearance. Cable (1974) has cautioned that the phylogenetic significance of stylets is uncertain, but as yet I see no reason to withdraw the suggestion. Indeed, further support can be found in the observations of Rothschild (1935) and Wright (1956) of a pharynx in the "sporocyst" of certain rhodometopous cercariae in contrast to Stunkard's (1964) failure to find any trace of a pharynx in the typically plagiorchioid sporocysts of a renicolid possessing a stylet cercaria.

* * *

I have not aimed at bibliographic completeness, nor have I ventured beyond my immediate interest into the realms of function and ultrastructure, important as these are to a full understanding, but have sought instead to point up the widespread occurrence of the system and something of its variable form and development in the hope that others will undertake studies of all aspects, including occurrence, ontogeny, ultrastructure, and function of both paranephridial and so-called lymphatic systems, and in so doing further illuminate the phylogeny of the Digenea.

Acknowledgments

In its present form, this essay owes much to three colleages. Professor Cable advised caution in too facile an acceptance of a common origin of all outgrowths. Dr Lester Cannon suggested the appendix as a way of clarifying and tightening the exposition of the thesis. Dr Robin Overstreet criticized both form and content trenchantly, forcing me to rethink and rewrite carefully. Cogency I owe to these three; the faults remaining are my own.

Permission was sought from R.N. Singh to use his 1959 figure (fig.4.3) from Studies on the morphology and life history of *Clinostomum piscidium* Southwell and Prashad 1918 (Trematoda:Clinostomatidae) in *Proceedings of the National Academy of Sciences, India, Sect. B* **29**:12–33, but no reply was received.

Permission was kindly granted by A. Fain and Académie Royale des Sciences d'Outre-Mer to use Fain's 1953 figure (fig.4.5) from Contribution à l'étude des formes larvaires des trématodes au Congo Belge et spécialement de la larve de *Schistosoma mansoni* in *Mémoires de l'Institut Royal Colonial Belge. Section des Sciences Naturelles et Médicales* 8° **22**: Fasc. 5, pp. 1–312.

References

Allison, L.N. 1943. *Leucochloridiomorpha constantiae* (Mueller) (Brachylaemidae), its life cycle and taxonomic relationships among digenetic trematodes. *Transactions of the American Microscopical Society* **62**:127–68.

Bartoli, P. 1967. Etude du cycle évolutif d'un trématode peu connu: *Lepocreadium pegorchis* (M. Stossich 1900) (Trematoda:Digenea). *Annales de Parasitologie humaine et comparée* **42**:605–19.

Blair, D. 1979. A new family of monostome flukes (Platyhelminthes:Digenea) from the dugong, *Dugong dugon* (Muller). *Annales de Parasitologie humaine et comparée* **54**:519–26.

Braun, M. 1893. Vermes. In *Bronn's Klassen und Ordnungen des Thier-Reichs.* Vol. 4, *Abteilung 1.* Leipzig: C.F. Winter.

Buttner, A. 1952. Cycle évolutif de *Ratzia joyeuxi* (E. Brumpt, 1922) (Trematoda: Opisthorchiidae). *Annales de Parasitologie humaine et comparée* **27**:105–42.

Cable, R.M. 1953. The life cycle of *Parvatrema borinqueñae* gen. et sp. nov. (Trematoda:Digenea) and the systematic position of the subfamily Gymnophallinae. *Journal of Parasitology* **39**: 408–21.

Cable, R.M. 1956. *Opistholebes diodontis* n.sp., its development in the final host, the affinities of some amphistomatous trematodes from marine fishes and the allocreadioid problem. *Parasitology* **46**:1–13.

Cable, R.M. 1974. Phylogeny and taxonomy of trematodes with reference to marine species. In *Symbiosis in the sea*, ed. W.B. Vernberg. Columbia: University of South Carolina Press.

Cable, R.M. 1982. Phylogeny and taxonomy of the malacobothrean flukes. In *Parasites — their world and ours. Proceedings of the fifth International Congress of Parasitology* Toronto, Canada 7–14 August 1982, ed. D.F. Mettrick and S.S. Desser, 194–97. Amsterdam, New York, Oxford: Elsevier Biomedical Press.

Cable, R.M. and Hunninen, A.V. 1942. Studies on *Deropristis inflata* (Molin), its life history and affinities to trematodes of the family Acanthocolpidae. *Biological Bulletin* **82**:292–312.

Cheng, Y.L. 1966. Comparative studies of the lymphatic system of four species of amphistomes. *Zeitschrift für Parasitenkunde* **27**:169–204.

Dickerman, E.E. 1934. Studies on the trematode family Azygiidae. I.The morphology and life cycle of *Proterometra macrostoma* Horsfall. *Transactions of the American Microscopical Society* **53**:8–21.

Fain, A. 1953. Contribution à l'étude des formes larvaires des trématodes au Congo Belge et spécialement de la larve de *Schistosoma mansoni. Mémoires de l'Institut Royal Colonial Belge. Section des Sciences Naturelles et Médicales*, 8° **22**: Fasc. 5, pp.1–312.

Feldman, S.I. 1941. Studies on the morphology and biology of a psilostome fluke. *Journal of Parasitology* **27**:525–33.

Freeman, R.F.H. and Llewellyn, J. 1958. An adult digenetic trematode from an invertebrate host: *Proctoeces subtenuis* (Linton) from the lamellibranch *Scrobicularia plana* (da Costa). *Journal of the Marine Biological Association of the United Kingdom* **37**:435–57.

Gibson, D.I. and Bray, R.A. 1979. The Hemiuroidea:Terminology, systematics and evolution. *Bulletin of the British Museum (Natural History) Zoological Series* **36**:35–146.

Ginetsinskaya, T.A. 1968. [Trematodes, their life cycles, biology and evolution] (in Russian). Leningrad: Academy of Sciences.

Grabda-Kazubska, B. 1980. *Euryhelmis zelleri* sp.n. and *Euryhelmis squamula* (Rudolphi, 1819) (Trematoda:Heterophyidae), metacercariae from *Rana temporaria* L., from the Babia Góra National Park, Poland. *Acta Parasitologica Polonica* **26**:115–28.

Hunninen, A.V. and Cable, R.M. 1943. The life history of *Lecithaster confusus* Odhner (Trematoda:Hemiuridae). *Journal of Parasitology* **29**:71–79.

Hussey, K.H. 1941. Comparative embryological development of the excretory system in digenetic trematodes. *Transactions of the American Microscopical Society* **60**:171–210.

Johnston, S.J. 1913. On some Queensland trematodes, with anatomical observations and descriptions of new species and genera. *Quarterly Journal of Microscopical Science* **59**:361–400.

Johnston, S.J. 1915. On *Moreauia mirabilis*, gen. et sp. nov., a remarkable trematode parasitic in *Ornithorhynchus. Proceedings of the Linnean Society of New South Wales* **40**:278–87.

Kawana, H. 1940. Study on the development of the excretory system of *Fasciola hepatica* L., with special reference of its first intermediate host in Central China. *Journal of the Shanghai Science Institute* Sect. IV **5**:13–34.

Køie, M. 1975. On the morphology and life history of *Opechona bacillaris* (Molin, 1859) Looss, 1907 (Trematoda:Lepocreadiidae). *Ophelia* **13**:63–86.

Komiya, Y. and Tajimi, T. 1941. Metacercariae from Chinese *Pseudorasbora parva* Temminck and Schlegel with special reference to their excretory system. I. (Metacercariae from Chinese fresh waters No. 1). *Journal of the Shanghai Science Institute* n.s. **1**:69–106.

Kuntz, R.E. 1952. Embryonic development of the excretory system in a pleurolophocercous (acanthostomatid) cercaria, three stylet cercariae (a microcercous cercaria, a brevicaudate, and a longicaudate dicrocoeliid cercaria) and in a microcaudate eucotylid cercaria. *Transactions of the American Microscopical Society* **71**:45–81.

Lang, B.Z. 1968. The life cycle of *Cephalogonimus americanus* Stafford, 1902 (Trematoda:Cephalogonimidae). *Journal of Parasitology* **54**:945–49.

La Rue, G.R. 1951. Host-parasite relations among the digenetic trematodes. *Journal of Parasitology* **37**:333–42.

La Rue, G.R. 1957. The classification of digenetic Trematoda: A review and a new system. *Experimental Parasitology* **6**:306–49.

Leão, A.T. 1944. Sistema excretor de *Renifer heterocoelium* (Travassos, 1921) Travassos, 1928 (Trematoda:Reniferinae). *Revista Brasileira de Biologia* **4**:109–12.

Linton, E. 1910. Helminth fauna of the Dry Tortugas. 2. Trematodes. *Papers from the Tortugas Laboratory (Department of Marine Biology)* **4**:11–98.

Looss, A. 1896. Recherche sur la faune parasitaire de l'Egypte. Première Partie. Mémoires de l'Institut Egyptien **3**:1–252.

Looss, A. 1899. Weitere Beiträge zur Kenntniss der Trematoden-Fauna Aegyptens, zugleich Versuch einer natürlichen Gliederung des genus *Distomum* Retzius. *Zoologische Jahrbücher. Abteilung für Systematik, Ökologie und Geographie der Tiere* **12**:521–784.

Looss, A. 1902. Ueber neue und bekannte Trematoden aus Seeschildkröten. *Zoologische Jahrbücher. Abteilung für Systematik, Ökologie und Geographie der Tiere* **16**:411–894.

Manter, H.W. 1935. The structure and taxonomic position of *Megasolena estrix* Linton 1910 (Trematoda) with notes on related trematodes. *Parasitology* **27**:431–39.

Manter, H.W. 1937. A new genus of distomes (Trematoda) with lymphatic vessels. *Reports on the Hancock Pacific Expeditions* **2**:11–22.

Manter, H.W. 1940. Digenetic trematodes of fishes from the Galapagos Islands and the neighbouring Pacific. *Reports on the Hancock Pacific Expeditions* **2**(14): 329–497.

Manter, H.W. 1947. Digenetic trematodes of marine fishes of Tortugas, Florida. *American Midland Naturalist* **38**: 257–416.

Ozaki, Y. 1937. Studies on the trematode families Gyliauchenidae and Opistholebitidae, with special reference to lymph system I. *Journal of Science of the Hiroshima University, Series B, Div. 1 (Zoology)* **5**:125–244.

Pearson, J.C. 1972. A phylogeny of life-cycle patterns of the Digenea. *Advances in Parasitology* **10**:153—89.

Price, E.W. 1932. The trematode parasites of marine mammals. *Proceedings of the United States National Museum* **81**:Art. 13, pp.1–68.

Reisinger, E. and Graack, B. 1962. Untersuchungen an *Codonocephalus* (Trematoda: *Strigeidae*), Nervensystem und paranephridialer Plexus. *Zeitschrift für Parasitenkunde* **22**:1–42.

Richard, J. 1971. La chétotaxie des cercaires. Valeur systématique et phylétique. *Mémoires du Muséum National d'Histoire Naturelle, Nouvelle Série. Série A, Zoologie* **67**:1–179.

Rothschild, M. 1935. The trematode parasites of *Turritella communis* Lmk. from Plymouth and Naples. *Parasitology* **27**:152–70.

Singh, R.N. 1959. Studies on the morphology and life history of *Clinostomum piscidium* Southwell and Prashad, 1918. (Trematoda:Clinostomatidae). *Proceedings of the National Academy of Sciences, India, Sect. B* **29**:12–33.

Skryabin, K.I. 1959. [Trematodes of animals and man.] Vol. 16 (in Russian). Moscow: Academy of Sciences.

Stunkard, H.W. 1943. A new trematode, *Dictyangium chelydrae* (Microscaphidiidae = Angiodictyidae), from the snapping turtle, *Chelydra serpentina. Journal of Parasitology* **29**:143–50.

Stunkard, H.W. 1964. Studies on the trematode genus Renicola: Observations on the life-history, specificity, and systematic position. *Biological Bulletin* **126**:467–89.

Stunkard, H.W. 1969. The morphology and life-history of *Neopechona pyriforme* (Linton, 1900) n.gen., n.comb. (Trematoda:Lepocreadiidae). *Biological Bulletin* **136**: 96–113.

Stunkard, H.W. and Cable, R.M. 1932. The life-history of *Parorchis avitus* (Linton), a trematode from the cloaca of the gull. *Biological Bulletin* **62**:328–38.

Ulmer, M. 1959. Studies on *Spirorchis haematobium* (Stunkard, 1922) Price, 1934 (Trematoda:Spirorchiidae) in the definitive host. *Transactions of the American Microscopical Society* **78**:81–89.

Watson, R.A. 1984. The life cycle and morphology of *Tetracerasta blepta*, gen. et sp. nov., and *Stegodexamine callista*, sp. nov., (Trematoda:Lepocreadiidae) from the long-finned eel, *Anguilla reinhardtii* Steindachner. *Australian Journal of Zoology* **32**:177–204.

Willey, C.H. 1935. The excretory system of the trematode, *Typhlocoelum cucumerinum*, with notes on lymph-like structures in the family Cyclocoelidae. *Journal of Morphology* **57**:461–71.

Willey, C.H. 1936. The morphology of the amphistome cercaria, *C. poconensis* Willey, 1930, from the snail, *Helisoma antrosa. Journal of Parasitology* **22**:68–75.

Willey, C.H. 1954. The relation of lymph and excretory systems in *Zygocotyle lunata. Anatomical Record* **120**:810–11.

Wilson, R.A. and Webster, L.A. 1974. Protonephridia. *Biological Reviews* **49**:127–60.

Wright, C.A. 1956. Studies on the life-history and ecology of the trematode genus *Renicola* Cohn, 1904. *Proceedings of the Zoological Society of London* **126**: 1–49.

Appendix

Families and representative genera showing a presumed paranephridial system

(see fig.4.7)

Line II = Hemiuroidea *sensu* Gibson & Bray 1979	
Azygiidae *Proterometra*	Dickerman 1934
Sclerodistomidae *Sclerodistomum*	Manter 1947; Gibson & Bray 1979
Hirudinellidae *Hirudinella*	Poirier
(as *Distomum clavatum*)	(fig. 1, pl. 32 in Braun 1893)
Botulus	Gibson & Bray 1979
Lecithasteridae *Lecithaster*	Hunninen & Cable 1943
Accacoeliidae *Mneiodhneria*	Monticelli (fig. 42 in Skryabin 1959)
Line III	
Fellodistomidae *Proctoeces*	Freeman & Llewellyn 1958
Gymnophallidae *Parvatrema*	Cable 1953
Brachylaemidae *Leucochloridiomorpha*	Allison 1943
Line IV	
Clinostomidae *Clinostomum*	Singh 1959
Cyathocotylidae	Sudarikov (in Skryabin 1959)
Diplostomidae	Sudarikov (in Skryabin 1959)
Strigeidae	Sudarikov (in Skryabin 1959)
Spirorchiidae *Spirorchis*	Ulmer 1959
Line V	
1. Notocotylidae *Catatropis* (as *Monostomum verrucosum*)	Looss 1896
Pronocephalidae *Diaschistorchis*	Johnston 1913
Rhabdiopoeidae *Rhabdiopoeus*	Johnston 1913
Opisthotrematidae *Opisthotrema*	Fischer (in Price 1932)
Labicolidae *Labicola*	Blair 1979,
Mesometridae *Mesometra* (as *Monostomum orbiculare*)	Parona (fig. 3, pl. 31 in Braun 1893)
2. Atractotrematidae *Atractotrema*	Pearson (unpublished)
3. Paramphistomidae *Paramphistomum*	Cheng 1966
Microscaphidiidae *Angiodictyum*	Looss 1902
4. Fasciolidae *Fasciola*	Querner (fig. 118 in Ginetsinskaya 1968); Kawana 1940
Cyclocoelidae *Typhlocoelum*	Willey 1935

Philophthalmidae *Parorchis*	Stunkard & Cable 1932
Echinostomatidae *Echinochasmus*	Komiya & Tajimi 1941
Psilostomidae *Psilostomum*	Feldman 1941
?Rhytidodidae *Rhytidodes*	Looss 1902
5. Heterophyidae *Euryhelmis*	Fraipont (fig. 8 pl. 30 in Braun 1893); Grabda-Kazubska 1980
Opisthorchiidae *Ratzia*	Buttner 1952
6. = Lepocreadioidea *sensu* Cable 1956	
With single outgrowth from bladder:	
Lepocreadiidae *Lepocreadium*	Bartoli 1967
Opechona	Køie 1975
Neopechona	Stunkard 1969; Køie 1975
With usually four blind "lymphatic" vessels:	
Homalometridae *Apocreadium*	Manter 1937
Megasolenidae *Megasolena*	Manter 1935
Hapladena	Manter 1947
Gyliauchenidae *Telotrema* (= *Gyliauchen*)	Ozaki 1937
7. Renicolidae *Renicola*	Wright 1956
Plagiorchiidae *sensu lato*	
Renifer	Leão 1944
Neorenifer	Pearson (unpublished)
Cephalogonimus	Lang 1968
Pachypsolus	Linton 1910
Enodiotrema	Looss 1902
Styphlodora	Looss 1899
Styphlotrema	Looss 1899
Unplaced	
Moreauiidae *Moreauia*	Johnston 1915

5 The Virulent Nature of *Babesia bovis*

I.G. Wright

Under field conditions, naturally transmitted *Babesia bovis* produces a hyperacute shock syndrome in susceptible cattle and as a result of its virulent nature *B. bovis* can produce very high rates of mortality. Other species of *Babesia* which infect cattle are generally much less virulent. Animals infected with *B. bovis* generally die with a parasitaemia of 1% or less whereas less virulent species such as *B. bigemina* produce parasitaemias of 10%–20% or more with only low mortality rates.

Studies were undertaken at Long Pocket Laboratories in 1973 to determine the nature of this virulence. It was found that *B. bovis* contained two proteases, one was active under acidic conditions, the other at neutral pH (Wright & Goodger 1973). These enzymes hydrolysed arginine esters and in addition hydrolysed fibrinogen and native haemoglobin. This observation helped to explain why *Babesia* parasites do not contain haemozoin, a pigment of incomplete hydrolysis of haemoglobin, which is a characteristic of malarial parasites. The hydrolysis of fibrinogen was also significant, for it was shown that disturbed metabolism of fibrinogen occurred in virulent *B. bovis* infections (Goodger 1975). This observation was extended by Goodger and Wright (1977) and Wright and Goodger (1977) who showed that cryofibrinogen complexes composed of altered fibrinogen and non-cross-linked fibrin were produced during severe *B. bovis* but not severe *B. bigemina* infections. These complexes were shown to adhere to infected erythrocyte surfaces and to capillary endothelium of small vessels, resulting in vascular stasis. This stasis in turn produced anoxic degenerative changes to tissues such as brain, kidney and skeletal muscle (Goodger, Wright & Mahoney 1981; Wright, Goodger, McKenna & Mahoney 1979). In addition to these changes Wright (1973), Wright and Mahoney (1974) and Wright (1977) showed that a marked involvement of the vasodilatory, hypotensive agents, kallikrein and kinin occurred from very early in the disease process when as few as 1×10^9 parasites were present in the blood stream. Both the alternate and classical complement pathways were also activated during acute babesiosis and terminally complete consumption of haemolytic complement occurred (Mahoney, Wright & Goodger 1980). The *in vivo* activation of plasma kallikrein by crude extracts of virulent parasites was demonstrated by Mahoney and Wright (1976), thus implicating the esterase in the pathogenesis of the disease. This extended an earlier observation by

Wright (1975) which showed that babesial esterase purified by affinity chromatography could directly activate plasma kallikrein *in vitro* and cause fibrinogen to form gel-like precipitates. Such properties were not demonstrable with extracts of avirulent strains of *B. bovis*.

The differences between virulent and avirulent strains of *Babesia* were demonstrated by a series of studies on irradiated and non-irradiated organisms. In the first of these Wright, Goodger and Mahoney (1980) showed that when organisms were irradiated with 35 Krads γ irradiation, they were attenuated. Although these organisms had the same multiplication rate and reached the same maximum parasitaemia as the non-irradiated virulent parent strain from which they were derived, all susceptible animals which received the irradiated organisms were only mildly affected whereas those that received the parent virulent strain had acute symptoms and died. Highly significant differences in changes to coagulation parameters, kinins and packed cell volume were observed in animals that received virulent organisms but not in these parameters in animals which received attenuated parasites.

In the second of these experiments differences in transmission rates by ticks between virulent and irradiated parasites were observed. *B. bovis* parasites which had been either freshly irradiated with 35 Krads γ radiation, or had been re-isolated from cattle infected 12 months previously with irradiated organisms were not transmitted transovarially by *Boophilus microplus*. Numbers of up to 1 x 10^5 larval progeny from replete female ticks which had fed on cattle with a patent parasitaemia of these irradiated organisms failed to transmit the organism to susceptible calves. Under similar conditions Mahoney and Mirre (1974) were able to transmit non-irradiated *B. bovis* with as few as 100 larvae. As a control experiment, Wright, Mirre, Mahoney and Goodger (1983) were able to transmit the parent non-irradiated strain of *B. bovis* after it had been in host cattle for 12 months. These experiments demonstrated two things.

Firstly, organisms attenuated by irradiation were not tick transmissble and this non-transmissible state was apparently permanent — at least for a minimum of 12 months. Secondly the non-transmissible state after a period of parasite recrudescence (Mahoney & Ross 1972) was not related to antigenic change; the control experiments reported by Wright, Mirre, Mahoney and Goodger (1983) demonstrated conclusively that virulent organisms could remain in carrier hosts for a minimum period of 12 months and yet still retain their infectivity for ticks. These data are in accord with Callow (1971) who reported that attenuation of organisms by blood passage through splenectomized calves could only be carried out for about 23 passages before organisms lost both their immunogenicity to cattle and their transmissibility by ticks. Both sets of data strongly suggest that avirulence and the ability to be transmitted by ticks are closely related.

However, in a further set of experiments Wright, Mahoney, Mirre, Goodger and Kerr (1982) have shown that avirulence and lack of immunogenicity are not necessarily interrelated. In a series of studies intact adult cattle were infected with *B. bovis* organisms freshly exposed to 35 Krads γ irradiation. From an innoculum containing 1 x 10^8 organisms, after irradiation an estimated dose of only 2 x 10^3 parasites remained. This residual population multiplied at the same rate as the non-irradiated virulent parent population, but had a delay of 5 days before it reached a similar maximum parasitaemia to that in control cattle which had received the parent strain. This delay was due to dose alone, the parasites multiplying tenfold per day. As the animals which had received irradiated organisms had only a mild disease it was concluded that those organisms which had survived the irradiation dose were a select avirulent population from the initial heterologous parent population. This selection was, moreover, stable. Parasites re-isolated from animals which had received a single dose of irradiated organisms 12 months previously were just as avirulent then as they were when freshly irradiated. Moreover, irradiated organisms stored in the vapour phase of liquid nitrogen also maintained a stable avirulent state. The most important finding from this work, however, was that animals immunized with avirulent, irradiated organisms were highly protected against challenge with virulent hererologous parasites. This immunity was as strong as that exhibited by naturally infected animals. Re-isolation of parasites from cattle that had received virulent organisms 12 months previously indicated that these parasites were as pathogenic as they had been prior to infection. This is in contrast to the work of Callow and Tammemagi (1967) who reported that strains recovered from long term carriers had become avirulent. However the recrudescence of *B. bovis* parasites in a host does not appear to affect either their virulence or immunogenicity. Mahoney, Wright and Goodger (1979) demonstrated that animals infected with avirulent and virulent strains of *B. bovis* were immune to virulent heterologous challenge four years after receiving the inoculum. The avirulent organisms in this case were produced by rapid passage through splenectomized calves; one of these strains was commercial blood vaccine as described by Callow (1971). The observations of Wright et. al. (1982) confirmed these findings, that avirulent organisms appeared to have very stable characteristics, and that both virulent and avirulent populations were as immunogenic as each other. Avirulence and immunogenicity are not mutally connected. Callow, Mellors and McGregor (1979) have reported that avirulent organisms regain their virulence after one passage through an intact animal. They hypothesized that the presence or absence of a spleen selects either virulent or avirulent organisms from an otherwise heterogeneous population. As this work has never been undertaken with cloned parasites the mechanism of this selection process remains unknown.

The nature of the virulence factor was determined by Wright, Goodger and Mahoney (1981) in avirulent populations produced either by irradiation or rapid passage through splenectomized calves. Both parent strain virulent parasites produced an acute, fatal disease after innoculation into susceptible cattle, with a marked involvement of blood kinins. The innoculation of the respective avirulent strains into susceptible cattle produced only a mild non-fatal disease, with no kinin involvement. Preparations of disrupted parasites were obtained from the four parasite populations. Both virulent strains contained high levels of esterase; the avirulent strains contained insignificant amounts. As all four strains had similar multiplication times it was concluded that the esterase was non-essential for the normal metabolic needs of the parasite in the vertebrate host, and that its presence was closely linked with pathogenic changes in the host. This was further confirmation of the earlier work which showed that parasite esterase directly activated the kallikrein-coagulation-complement cascade. As parasites which lack esterase are also non-tick transmitted, it raises the question as to whether the enzyme is necessary for the initial infection of the tick gut cells by merozoites. Conclusive evidence for the role of esterase in the invation of tick gut cells by merozoites is as yet not forthcoming.

Recent work at our laboratories has further characterized this esterase. The enzyme was purified from a crude extract of *B. bovis* by affinity chromatography using Soya Bean Trypsin inhibitor as a ligand. In native form this enzyme had a molecular weight > 200 x 10^3, but on denaturing gels major bands were observed with molecular weights of 20 x 10^3, 10 x 10^3 and 7 x 10^3 (fig. 5.1). Western transfer analysis using bovine anti-*B. bovis* esterase serum revealed a major band with a molecular weight of 19–20 x 10^3 (fig. 5.2). Using indirect immunofluorescence, both bovine and rabbit antisera to the esterase avidly stained infected red blood cells with only weak parasite staining (fig. 5.3). This immunofluorescent data strongly suggest that the esterase is secreted by the parasite during its multiplicative phase in the infected red cell. The reason for this secretion is unknown, but immunofluorescence suggests that larger amounts of esterase are present in red cells which contain recently divided parasites. Presumably the esterase is released into the host's circulation when recently divided parasites rupture and emerge from the red cell. These enzymes cannot be involved in the rupture of the red cell for avirulent organisms lacking in esterase undergo the process as readily as virulent ones. This probably release of esterase into the circulation would then activate the kallikrein-coagulation-complement cascade.

The potential protective effect of antibodies against this enzyme was investigated using both virulent homologous and heterologous challenge of animals previously immunized with affinity-purified extracts of this esterase (Wright, Goodger, Rode-Bramanis, Mattick, Mahoney & Waltisbuhl, 1983). Limited protection was afforded against homologous challenge, but no protection was observed in animals undergoing heterologous challenge. These data point strongly to the strain-specific nature of the enzyme. A high degree of esterase inhibition was demonstrated with sera from the homologous immunized animals and the

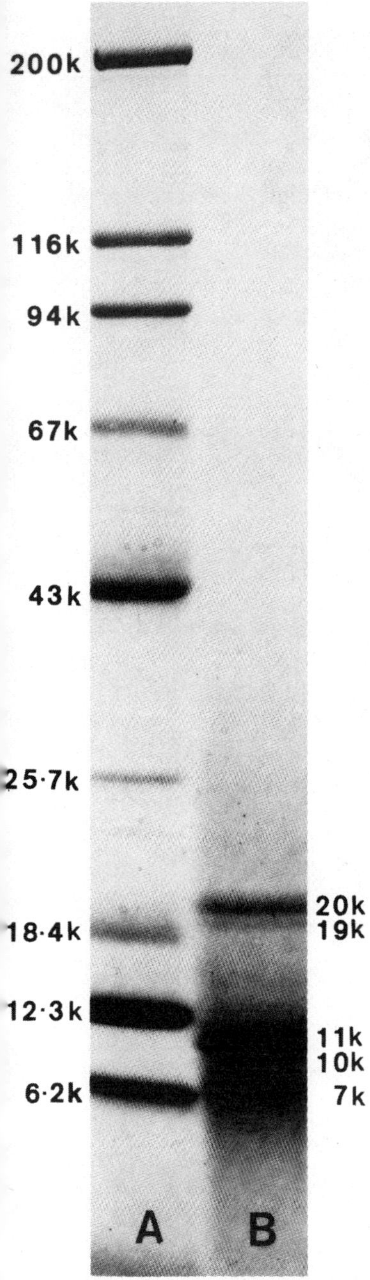

gure 5.1 *Lane A*. Gradient acrylamide electro-
oresis under denaturing conditions of protein
ndards. *Lane B*. Affinity-purified *B. bovis*
erase preparation. (From *Z. Parasitenkunde*
).

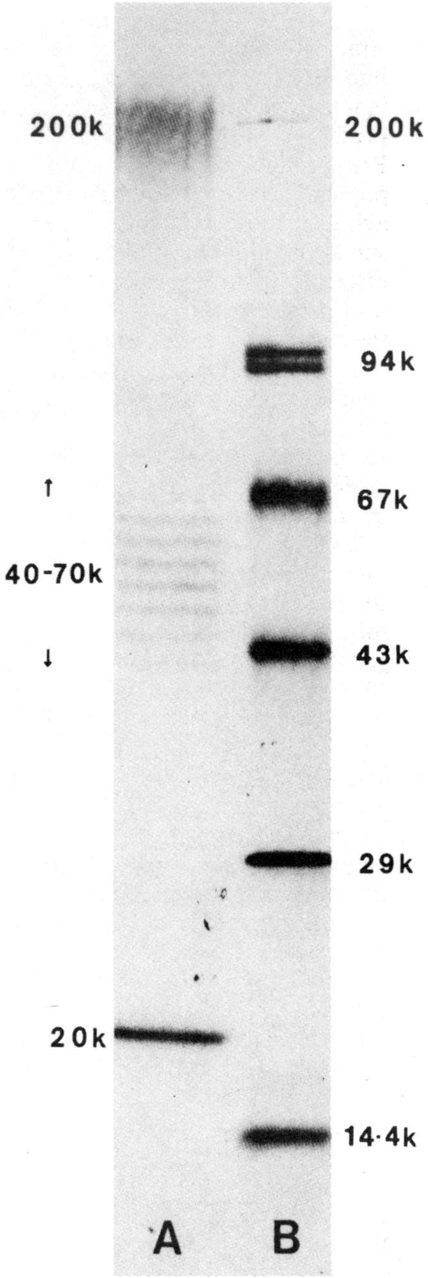

Figure 5.2 *Lane A*. Western transfer analysis of *B. bovis* esterase preparation using bovine anti *B. bovis* esterase sera. *Lane B*. ^{14}C-labelled protein standards. (From *Z. Parasitenkunde* **69**).

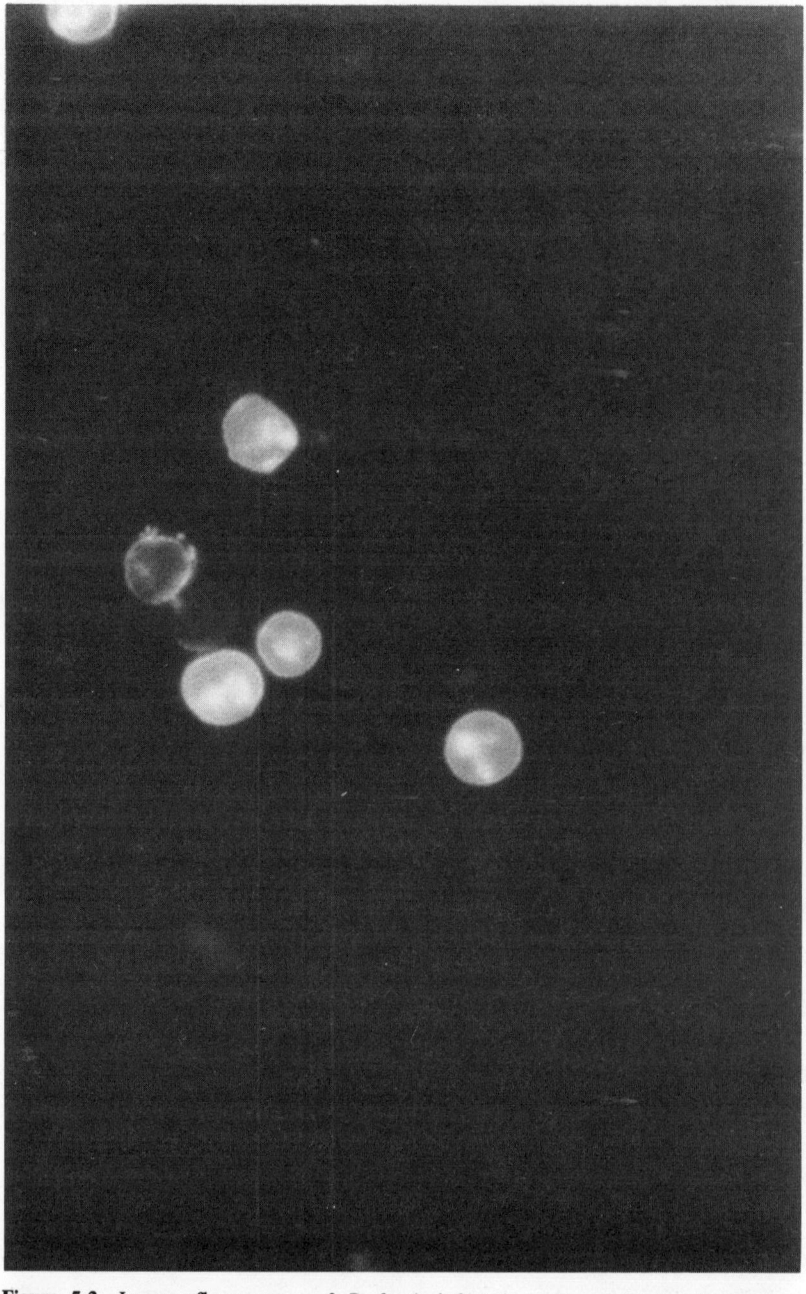

Figure 5.3 Immunofluorescence of *B. bovis* infected erythrocytes using fluorescein isothiocyanate-labelled anti-*B. bovis* esterase sera.

activity of the sera declined significantly in terminal infections, indicating that antibodies do play a part in the removal of esterase from the circulation in homologous but not heterologous infections. These antibodies precipitate the antigen and may form complex molecules with cryofibrinogen, conglutinin, $\alpha 2$ macroglobulin, fibronectin and other proteins of acute inflammation, but during infection the amount of esterase apparently overwhelms such defences.

The nature of *B. bovis* virulence is now more clearly understood. This virulence is closely related to the presence or absence of an esterase which is not essential for the parasites' well-being in the vertebrate host. The presence or absence of this esterase has no relationship to the immunogenicity of the parasite strain. However, it is possible that this enzyme may play a role in the transmission of the parasite by its tick vector. The demonstration of such a role may lead to an important vaccine source as a means of reducing transmission of the organism in the environment. The protection against homologous challenge and the strain-specific nature of the esterase may explain why relapses of parasitaemia of *B. bovis* are subclinical as only the homologous system is involved. In heterologous challenge conditions clinical responses are observed, presumably because immunity to the strain-specific esterase is absent. Protective immunity is not dependent on this antigen or on the degree of parasite virulence.

Acknowledgments

Permission to reproduce figures 5.1 and 5.2 from *Zeitschrift für Parasitenkunde* **69**:708–9 (1983) was kindly granted by Springer-Verlag, Heidelberg, publishers of the journal.

References

Callow, L.L. 1971. The control of babesiosis with a highly infective, attentuated vaccine. *Proceedings of the 19th World Veterinary Congress* **1**:357–59.

Callow, L.L., Mellors, L.T. and McGregor, W. 1979. Reduction in virulence of *Babesia bovis* due to rapid passage in splenectomized cattle. *International Journal for Parasitology* **9**:333–38.

Callow, L.L. and Tammemagi, L. 1967. Vaccination against bovine babesiosis. Infectivity and virulence of blood from animals either recovered from or reacting to *Babesia argentina. Australian Veterinary Journal* **43**:249–56.

Goodger, B.V. 1975. A cold precipitable fibrinogen complex in the plasma of cattle dying from infection with *Babesia argentina. Zeitschrift für Parasitenkunde* **48**:1–7.

Goodger, B.V. and Wright, I.G. 1977. *Babesia bovis (argentina)*: Observations of coagulation parameters, fibrinogen catabolism and fibrinolysis in intact and splenectomised cattle. *Zeitschrift fur Parasitenkunde* **54**:9–27.

Goodger, B.V., Wright, I.G. and Mahoney, D.F. 1981. Changes in conglutinin, immunoconglutinin, complement C_3 and fibronectin concentrations in cattle acutely infected with *Babesia bovis. Australian Journal of Experimental Biology and Medical Science* **59**:531–38.

Mahoney, D.F. and Mirre, G.B. 1974. *Babesia argentina*: The infection of splenectomized calves with extracts of larval ticks (*Boophilus microplus*). *Research in Veterinary Science* **16**:112–14.

Mahoney, D.F. and Ross, D.R. 1972. Epizootological factors in the control of bovine babesiosis. *Australian Veterinary Journal* **48**:292–98.

Mahoney, D.F. and Wright, I.G. 1976. *Babesia argentina*: Immunization of cattle with a killed antigen against infection with a heterologous strain. *Veterinary Parasitology* **2**:273–82.

Mahoney, D.F., Wright, I.G. and Goodger, B.V. 1979. Immunity in cattle to *Babesia bovis* after single infections with parasites of various origin. *Australian Veterinary Journal* **55**:10–12.

Mahoney, D.F., Wright, I.G. and Goodger, B.V. 1980. Changes in the haemolytic activity of serum complement during acute *Babesia bovis* infection in cattle. *Zeitschrift für Parasitenkunde* **62**:39–45.

Wright. I.G. 1973. Plasma kallikrein levels in acute *Babesia argentina* infections in splenectomised and intact calves. *Zeitschrift für Parasitenkunde* **41**:269–80.

Wright, I.G. 1975. The probable role of *Babesia argentina* esterase in the *in vitro* activation of plasma prekallikrein. *Veterinary Parasitology* **1**:91–96.

Wright, I.G. 1977. Kinin, kininogen and kininase levels during acute *Babesia bovis* (= *B. argentina*) infection of cattle. *British Journal of Pharmacology* **61**:567–72.

Wright, I.G. and Goodger, B.V. 1973. Proteolytic enzyme activity in the intra-erythrocytic parasites *Babesia argentina* and *Babesia bigemina*. *Zeitschrift für Parasitenkunde* **42**:213–20.

Wright, I.G. and Goodger, B.V. 1977. Acute *Babesia bigemina* infection: Changes in coagulation and kallikrein parameters. *Zeitschrift für Parasitenkunde* **53**:63–73.

Wright, I.G., Goodger, B.V. and Mahoney, D.F. 1980. The irradiation of *Babesia bovis*. I. The difference in pathogenicity between irradiated and non-irradiated populations. *Zeitschrift für Parasitenkunde* **63**:47–57.

Wright, I.G., Goodger, B.V. and Mahoney, D.F. 1981. Virulent and avirulent strains of *Babesia bovis*: The relationship between parasite protease content and pathophysiological effect of the strain. *Journal of Protozoology* **28**:118–20.

Wright, I.G., Goodger, B.V., McKenna, R.V. and Mahoney, D.F. 1979. Acute *Babesia bovis* infection: A study of the vascular lesions in kidney and lung. *Zeitschrift für Parasitenkunde* **60**:19–27.

Wright, I.G., Goodger, B.V., Rode-Bramanis, K., Mattick, J.S., Mahoney, D.F. and Waltisbuhl, D.J. 1983. The characterisation of an esterase derived from *Babesia bovis* and its use as a vaccine. *Zeitschrift für Parasitenkunde* **69**:703–14.

Wright, I.G. and Mahoney, D.F. 1974. The activation of kallikrein in acute *Babesia argentina* infections of splenectomised calves. *Zeitschrift für Parasitenkunde* **43**:271–78.

Wright, I.G., Mahoney, D.F., Mirre, G.B., Goodger, B.V. and Kerr, J.D. 1982. The irradiation of *Babesia bovis*. II. The immunogenicity of irradiated blood parasites for intact cattle and splenectomised calves. *Veterinary Immunology and Immunopathology* **3**:591–601.

Wright, I.G., Mirre, G.B., Mahoney, D.F. and Goodger, B.V. 1983. Failure of *Boophilus microplus* to transmit irradiated *Babesia bovis*.

Part Three

Hosts

6 Immunological Responses of Mammals to Ectoparasites: Mosquitoes and Ticks

J.R. Allen

Introduction

Mosquitoes and Ixodid ticks are infamous as vectors of microorganisms pathogenic for man and domestic animals, and there is much published information on the transmission of such microorganisms and the hosts' immunological reactions to them. There is relatively little known about mammals' immunological reactions to the ectoparasites themselves, reactions to the uninfected mosquito or tick. In this review, immunological reactions to bites of the rapidly feeding mosquito will be considered. These reactions are irritating to the host and in some cases life-threatening, but appear to cause no direct damage to the insect. Then, by way of comparison, the similar immunological reactions to the more slowly-feeding ixodid ticks will be discussed, reactions which do appear to protect hosts and produce deleterious effects on the parasites.

Immunological Reactions to Mosquito Bites

Skin reactions to mosquito bites are experienced commonly by most individuals. These reactions were thought to be caused by components of the mosquito saliva which was shown by Gordon and Lumsden (1939) to be deposited in the dermis during the skin-probing and rapid blood feeding activities of female mosquitoes. Hudson, Bowman and Orr (1960) confirmed this. They were able to cut the common salivary duct of *Aedes stimulans* and to show that these mosquitoes could blood feed after the operation. If normal and treated mosquitoes fed on the same human volunteer, normal mosquito bites elicited skin reactions, but bites from the treated mosquitoes did not.

Early workers debated whether the skin reactive components of the mosquito's saliva were directly toxic or antigenic. McKiel and West (1961) presented evidence to support the suggestion that the skin reactions represent immunological (hypersensitivity) reactions. There are 2 types of skin reactions to mosquito bites in man. The early or immediate reaction develops a few minutes after the bite, and consists of a wheal and surrounding flare, accompanied by pruritus. This type of reaction usually disappears within 60–90 min. The second type of reaction is delayed in appearance, developing several hours after the bite and reaching a maximum 24–48 h thereafter. This reaction consists of a raised red indurated area accompanied by severe pruritus.

The first line of evidence supporting the suggestion that these skin reactions represent hypersensitivity reactions concerns the sequence of reactions in individuals subjected to repeated mosquito bites. Mellanby (1946) in Britain, assembled volunteers who would have been unlikely to have been bitten previously by *Aedes aegypti*, an exotic species. Volunteers were repeatedly bitten by the mosquitoes over periods of several months. Heilesen (1949) performed similar experiments using very young children. The sequence of skin reactions occurring in these experiments involved five stages. No reactions were elicited by the first few bites (stage 1). Delayed reactions to bites occurred in stage 2. Later in the series, a single bite would elicit both an immediate and a delayed reaction (stage 3). Later again (stage 4) only an immediate response occurred. Mellanby (1946) surmised that most individuals would eventually become non-reactive (stage 5). Laboratory animals subjected to repeated mosquito bites also exhibited this sequence of skin reactions (McKiel & West 1961).

The same sequence of skin reactions was observed by Mote and Jones (1936) when they repeatedly injected small doses of foreign protein intradermally into human volunteers. The delayed reactions produced in this way differed from the classical Type IV delayed hypersensitivity reactions and were labelled Jones-Mote reactions. More recently Richerson, Dvorak and Leskowitz (1969) reinvestigated the Jones-Mote hypersensitivity reaction and showed that marked infiltrations of basophil leukocytes were characteristic of these reactions in guinea pigs and other mammals. Such reactions were renamed cutaneous basophil hypersensitivity reactions.

Histopathological studies of immediate skin reactions to mosquito bites in guinea pigs were reported by French (1972). He observed marked infiltrations of eosinophils, which would be consistent with the suggestion that these reactions represented immediate hypersensitivity reactions. He also reported infiltrations of neutrophils into reaction sites 60–90 min after the bite, which suggested to him that Arthus-like reactions may also occur. No attempts to demonstrate basophil leukocytes in the reactions to mosquito bites have been reported.

The second line of evidence supporting the hypersensitivity hypothesis comes from attempts to sensitize laboratory animals artificially, with mosquito antigens, so that they would react to their first mosquito bite. McKiel (1959) successfully sensitized laboratory animals using antigenic extracts of mosquito tissues. In other experiments, oral secretion was collected from *A. aegypti* mosquitoes by persuading them to feed through artificial membranes on distilled water. Salivary components injected into the water by the mosquitoes were freeze dried and used subsequently to sensitize guinea pigs (Allen & West 1966). Newsome, Jones, French and West (1969) isolated the skin reactive component of oral secretion and found it to be a high molecular weight protein with an amino acid composition not unlike that of several invertebrate collagens.

The third line of evidence comes from experiments designed to demonstrate passive transfers of hypersensitivity to mosquito bites. Passive transfers of immediate hypersensitivity to mosquito bites with serum were achieved in humans by Brown, Griffitts, Erwin and Dyrenforth (1938) and Rockwell and Johnson (1952), using essentially the Prausnitz-Küstner technique. Attempts to demonstrate passive transfers between experimental animals have been made. No success was reported using serum or lymphoid cells from animals which had been sensitized by mosquito bites, but serum transfers of immediate hypersensitivity to bites, and cell transfers of delayed hypersensitivity were obtained if serum or cells were taken from guinea pigs which had been artificially sensitized with oral secretion and adjuvant (Allen 1966).

This rather fragmentary collection of evidence has persuaded many that skin reactions to mosquito bites do represent local hypersensitivity reactions. These immunological reactions are irritating to the host, and sometimes much more serious. Some individuals exhibit severe skin reactions to the bites. Brown et al. (1938) and Tager, Lass, Gold and Lengy (1969) for example, reported severe necrotic skin lesions in mosquito-bitten patients. Severe systemic reactions to mosquito bites in man have also been reported, both systemic anaphylactic responses and severe delayed systemic reactions such as those described by Suzuki, Negishi, Tomizawa, Shibasaki, Kuroume and Matsumura (1976).

In human cases with severe reactions, attempts have been made to desensitize or hyposensitize the patients. Relatively very crude antigenic extracts of insects have been used in these procedures in contrast to the highly purified antigens used successfully in the desensitizations of patients suffering from bee- and wasp-sting allergies, for example. It would seem entirely possible, with current biotechnology, to obtain purified antigens for use in mosquito bite allergies. Although it is known that the antigens in the saliva of one species of mosquito are not identical with those from other species, there are apparently shared antigens, since there are two reports by Tager et al. (1969) and Rockwell and Johnson (1952) of success in desensitizing individuals hypersensitive to *Culex pipiens* bites using antigens derived from *A. aegypti*.

The immunological reactions to mosquito bites may thus be considered to be non-protective and, if anything, deleterious to the host. Female mosquitoes apparently feed equally well on naive and sensitized hosts, and their progeny is not adversely affected by feeding on a sensitized host. If, however, mosquitoes feed on hosts immunized with mosquito antigens which are not introduced by the normally feeding mosquito, those mosquitoes may show some deleterious effects following the bloodmeal. Alger and Cabrera (1972) showed increased death rates in *Anopheles stephensi* females which had fed on rabbits immunized with mosquito midgut antigens. And, in similar experiments, Sutherland and Ewen (1974) observed substantially reduced fecundity of *A. aegypti* which fed on rabbits or guinea pigs immunized with whole extracts of these mosquitoes. In both these reports the authors speculated that the effects on mosquitoes were antibody-mediated.

Even though the immunological reactions to mosquito bites are non-protective in that they normally produce no deleterious effects on the mosquito, there is some evidence to suggest that immunological reactions to mosquito salivary antigens may protect hosts against viruses and protozoan organisms transmitted by mosquitoes. Feinsod, Spielman and Waner (1975), working with the mosquito-transmitted Sindbis virus, showed that sera from guinea pigs sensitized by bites of uninfected mosquitoes neutralized virus which had been propagated in mosquito cells, but not virus propagated in mammalian cells. These results could be explained by mosquito antigens being incorporated in the virion envelope in the former case, these antigens acting as targets for the anti-mosquito antibodies. Alger, Harant, Willis and Jorgensen (1972) studied *Plasmodium berghei* infections in mice. Mice which had been immunized with salivary glands from uninfected mosquitoes were significantly protected against intraperitoneal challenge with *P. berghei* sporozoites. Again, the authors suggested that mosquito antigens incorporated in the sporozoite membranes might act as targets for the anti-mosquito antibodies. These suggestions deserve further investigation.

Immunological Reactions to the Bites of Ixodid Ticks

Ixodid ticks also introduce salivary components into the skin of their hosts as they feed. Gregson (1967) has demonstrated the repeated injection of salivary secretions during the feeding process of *Dermacentor andersoni*, for example. However, unlike mosquitoes, Ixodid ticks feed for prolonged periods while attached to the host skin. Larval, nymphal and adult stages may remain attached and feeding for periods of 5–15 days. Some mammalian hosts develop a resistance to tick feeding, and almost all the studies of immunological reactions to tick bites have been done in relation to this tick resistance phenomenon.

Tick resistance is manifested by an inability of most challenging ticks to feed normally on the resistant host. It is acquired by cattle and laboratory animals, but apparently not by all mammalian hosts. Theis and Budwiser (1974), for example, found no evidence of its occurrence in dogs infested with *Rhipicephalus sanguineus*. As Willadsen (1980) states, despite the increasing number of tick-host systems investigated in recent times, the mechanisms of tick resistance in any one system are still incompletely understood. For comprehensive reviews of this subject, those of Willadsen (1980) and Wikel (1982) are recommended. In this section, the immunological responses to Ixodid ticks in guinea pigs and cattle will be discussed, with the intention of comparing them to the immunological responses elicited by mosquito bites and of suggesting possible mechanisms of tick resistance.

A. Tick resistance in guinea pigs

Trager (1939) was the first to suggest that immunological mechanisms were responsible for tick resistance. Working with guinea pigs, he observed that larvae of *Dermacentor* species engorged less readily on previously infested than on naive animals, and provided some evidence that the resistant state could be passively transferred with serum from resistant donors.

The guinea pig has since been used by several workers in experiments with several tick species, and though the guinea pig is not the natural host for larvae of the tick species investigated, some intriguing information has emerged. Immunological mechanisms are involved in the guinea pig's tick resistance. Allen (1973) and B.G. Bagnall (unpublished Ph.D. thesis, University of Sydney, 1975) confirmed that resistance was acquired. The larvae of *D. andersoni* and *Ixodes holocyclus* were found to feed less readily on guinea pigs which had already experienced tick infestations, and this has been observed also by several authors using guinea pigs and other species of ticks (Heller-Haupt, Varma & Langi 1981; Askenase, Bagnall & Worms 1982).

The administration of immunosuppressive drugs, during primary infestations with larvae of *D. andersoni*, prevented guinea pigs from acquiring resistance to a subsequent larval infestation (Allen 1973; Wikel & Allen 1976a).

Histopathological studies of tick-infested skin from susceptible and resistant guinea pigs have revealed that heavy infiltrations of basophil leucocytes occur close to the attachment sites of larvae on resistant animals. Allen (1973) observed very large numbers of basophils in the dermis and particularly in epidermal vesicles beneath the sites of attachment of *D. andersoni* larvae. Most of these cells were at least partially degranulated. Bagnall (1975, thesis cited above) reported similar findings in resistant guinea pigs challenged with *I. holocyclus* larvae. Resistance to other Ixodid tick species has also been found to be associated with cutaneous basophil reactions in guinea pigs (Brown & Askenase 1981; Askenase, Bagnall & Worms 1982; Krinsky, Brown & Askenase 1982). It appears likely that such reactions represent cutaneous basophil hypersensitivity reactions which are homologous with the "Jones-Mote" skin reactions observed in responses to mosquito bites.

Attempts to immunize guinea pigs artifically with tick-derived antigens have been successful, inducing a degree of resistance to the first tick challenge in immunized hosts. Trager (1939) obtained evidence of a degree of resistance to larval challenge in guinea pigs immunized with a larval extract. Bagnall (1975, thesis cited above) protected guinea pigs against larvae of *Ixodes holocyclus*, also using an extract of whole larvae. Wikel (1981a) used an extract of salivary glands from partially fed female *D. andersoni* to induce protection in guinea pigs against larval challenge. Allen and Humphreys (1979) used extracts of internal organs of partially fed female *D. andersoni* and induced significant resistance to the feeding of adult ticks in guinea pigs.

Further support for the immunological mediation of tick resistance in guinea pigs comes from experiments designed to reveal passive transfers of resistance. Trager (1939) observed the possibility that serum transfers might confer a degree of resistance on recipients of serum. Bagnall (1975, thesis cited above) demonstrated significant transfer of resistance to *I. holocyclus* between syngeneic guinea pigs using lymph node cells from resistant donors, and an additive effect was noted when both serum and cells were transferred. Adoptive transfer of resistance to *D. andersoni* was obtained by Wikel and Allen (1976b) using viable lymph node cells. They found that passive transfer of serum resulted in reduced weights of larvae feeding on serum recipients. Askenase, Bagnall and Worms (1982) showed passive transfers of resistance and cutaneous basophil hypersensitivity responses to *I. holocyclus* and *Rhipicephalus appendiculatus* with cells or serum from resistant guinea pigs. From such results one might conclude that both cellular and humoral immune mechanisms contribute to tick resistance in guinea pigs.

Correlates of cell mediated responses to tick salivary gland antigens were demonstrated in resistant guinea pigs by Wikel, Graham and Allen (1978). Both delayed skin reactions and *in vitro* blastogenesis of lymph node cells were elicited by salivary gland antigens from *D. andersoni*. Also, sera from resistant guinea pigs have been shown to contain tick-specific antibodies, both precipitating and homocytotropic (Bagnall 1975, thesis cited above).

Thus, whatever the guinea pig's mechanism of tick resistance may be, it appears that cell mediated and humoral immune components are involved, and one might expect that cutaneous basophil hypersensitivity reactions, which occur locally at the sites where challenging ticks attach to the resistant animals, may be of some importance.

B. Tick resistance in cattle

Studies of the reactions of cattle to *Boophilus microplus* by scientists in Australia have provided most of the information on tick resistance in cattle. Work by Roberts (1968) with *Bos taurus* and by Wagland (1975) with *Bos indicus* has confirmed that the state of resistance is acquired, not innate. Resistance was not evident at the time of primary infestations, but became evident only after the hosts had received one or more infestations. With *B. microplus*, a one-host tick, highly resistant animals allow less than 1% of the infesting larvae to develop through and engorge as adults. High level resistance tends to be acquired more readily by *Bos indicus* and their crosses, but some *Bos taurus* individuals and their progeny also show this ability (Hewetson 1971; Utech, Seifert & Wharton 1978).

Acquired resistance to *B. microplus* in cattle, like tick resistance in guinea pigs, appears to be an immunological phenomenon. Riek (1962) suggested that resistance was associated with a hypersensitivity to salivary secretions of the tick. Histological studies of skin reactions to *B. microplus* infestations of susceptible and resistant cattle seem consistent with this suggestion. Tatchell and Moorhouse (1968) and Moorhouse and Tatchell (1969) provided detailed histological studies showing degranulation of mast cells, infiltration of eosinophil leucocytes and epidermal vesiculation in varying degrees in the infested skin of susceptible and resistant animals. Schleger, Lincoln, McKenna, Kemp and Roberts (1976) studied early lesions in *Bos taurus* and observed that these reactions were significantly greater in highly resistant animals. Histological studies of bovine skin reactions to adult *I. holocyclus* ticks were performed on naive and previously infested hosts (Allen, Doube & Kemp 1977). Epidermal vesiculation, eosinophil infiltrations and the presence of basophil leucocytes in the dermis and epidermis were reported in the challenged skin of previously infested animals. The basophil response in this case was, however, much less marked than that shown in tick resistant guinea pigs.

Roberts and Kerr (1976) showed that transfer of plasma from cattle highly resistant to *B. microplus* conferred a significant degree of resistance on recipient calves. Relatively very large volumes, 40 ml plasma/kg, were used in these experiments. Plasma from cattle exhibiting low levels of tick resistance was ineffective. Adoptive transfers of tick resistance between cattle have apparently not been attempted.

Artificial immunizations of cattle with crude antigenic extracts of tick tissues have induced significant but not dramatic tick resistance. Partial resistance was obtained by Brossard (1976) in cattle immunized with salivary gland extracts of *B. microplus*. Extracts of midgut and reproductive organs (not salivary glands) from partially fed female *D. andersoni*, with Freund's incomplete adjuvant, were found to induce significant resistance in cattle to adult *D. andersoni* infestations (Allen & Humphreys 1979). The weights of female ticks feeding on immunized animals were reduced as were the ticks' progeny. It is possible that the use of non-salivary antigens in this work induced immunological responses different from those induced in naturally acquired tick-resistance. It was suggested that antibodies specific for tick midgut or other tissues might be ingested by challenging ticks, with resulting deleterious effects on the ticks. This situation could be analogous to the previously mentioned findings of Alger and Cabrera (1972) and Sutherland and Ewen (1974) in their experimental immunizations of rabbits with midgut and other tissues of mosquitoes. In this connection, it has been shown that antibodies ingested by *D. variabilis* can pass across the digestive tract and into the haemolymph of the tick, and react specifically with tick antigens against which the antibodies were raised (Ackerman, Clare, McGill & Sonenshine 1981).

Very crude antigenic extracts were used in these artifical immunizations of cattle. It is likely that pure appropriate tick antigens can be produced by current biotechnological methods and used to induce high levels of tick resistance in domestic animals.

Mechanisms of Tick Resistance

It is probable that several immunologically mediated mechanisms are at work in tick resistance, but one plausible mechanism common to both cattle and guinea pigs rests on the suggestion that the local release of histamine or other mediators from degranulating basophils or mast cells in the skin of resistant animals disrupts the normal feeding behaviour of the ticks.

In guinea pigs, as previously mentioned, large numbers of degranulated basophils have been shown to occur in the skin of challenged resistant guinea pigs. Brown, Galli, Gleich and Askenase (1982) have shown that the cutaneous basophil response in guinea pigs is a requirement for expression of tick resistance in that treatment of guinea pigs with anti-basophil serum ablated both the basophil response and the expression of resistance.

Degranulated mast cells and basophils have been demonstrated also in the skin of challenged resistant cattle. Moreover, Willadsen, Williams, Roberts and Kerr (1978) showed that immediate skin reactions elicited in cattle by purified antigenic fractions from *B. microplus* correlated with the degree of tick resistance exhibited by individual animals. These results also are consistent with the suggestion that immediate hypersensitivity reactions in the skin, resulting in the degranulation of mast cells or basophils, may play a role in tick resistance.

Histamine concentrations in the skin of tick-challenged cattle and guinea pigs were found to be significantly higher in resistant than in susceptible animals (Willadsen, Wood & Riding 1979; Wikel 1981b). In resistant guinea pigs, concentrations of histamine as high as 10^{-2}M have been found in epidermal vesicles which developed beneath attached larvae of *D. andersoni* (Paine, Kemp & Allen 1983).

Attached ticks could be affected in several ways by the local release of histamine and other mediators from mast cells or basophils. The release of such mediators causes pruritus, and this together with reflex grooming activity of the host could disrupt the ticks' normal feeding behaviour. Koudstaal, Kemp and Kerr (1978) showed that the extent of grooming activities in cattle was correlated with the degree of tick resistance of the different animals, and estimated that 50% of *B. microplus* larvae infesting resistant animals might be removed by grooming.

Local mediator release also appears to affect ticks, more directly, in other ways. Kemp and Bourne (1980) observed that the introduction of histamine into the feeding medium of *in vitro* fed *B. microplus* larvae caused them to detach from artificial skin membranes. The addition of serotonin, bradykinin or prostaglandin E_2 to the feeding medium did not produce this effect.

The feeding and salivation activities of *D. andersoni* adults have also been shown to be reduced by adding histamine and serotonin to the tick's feeding medium (Paine et al. 1983). Ticks fed *in vitro*, attached to artificial membranes, were incorporated into electronic circuits which recorded the oscillating resistance changes associated with sucking and salivation. The amplitudes of these recordings were significantly reduced, often almost to zero, when 10^{-2}M concentrations of histamine and serotonin were added to the feeding medium. Additions of dopamine, prostaglandin E_2 and $F_{2\alpha}$ did not produce this effect.

The results from *in vitro* experiments correlate with observations made on the behaviour of ticks infesting resistant hosts. In cattle which were prevented from grooming, Kemp, Koudstaal, Roberts and Kerr (1976) observed that *B. microplus* larvae, during the first 16 h of infestations, spent significantly less time feeding and more time wandering on resistant animals. Similarly, on guinea pigs prevented from grooming, *D. andersoni* larvae detached or changed attachment sites significantly more often on resistant animals (Allen & Kemp 1982).

There is, thus, considerable circumstantial evidence to support the suggestion that hypersensitivity reactions, involving basophils and mast cells, disrupt the feeding behaviour of Ixodid ticks on resistant hosts, just as there is evidence to suggest that similar reactions may deleteriously affect nematode endoparasites (Askenase 1980). Exactly how the locally released mediators directly or indirectly cause changes in tick behaviour awaits elucidation.

Another type of cell, which also seems to have some role in tick resistance mechanisms, is the Langerhans cell. This dendritic cell, which occurs in the epidermis and other epithelia of mammals, has been shown to play a part in immunological reactions of the skin, including contact hypersensitivity reactions. Stingl (1980), Friedmann (1981) and others have reviewed the properties and proposed functions of Langerhans cells. Briefly, these cells make up 3%–8% of the cell population in the epidermis and possess several properties in common with Ia-bearing macrophages. Among their proposed functions are the ability to trap antigens or haptens passing through the epidermis and the ability to move from the epidermis and present antigens to relevant lymphocytes in the dermis or draining lymph nodes. Apart from this role in the afferent limb of immunological responses, there is reason to suggest that antigen-laden Langerhans cells, in the epidermis of immunized animals, act as target cells which provide an early focus for humoral and cellular immune reactions in the skin (Silberberg-Sinakin, Fedorko, Baer, Rosenthal, Berezowsky and Thorbecke, 1977; Stingl 1980).

In studies designed to reveal the location of tick salivary antigens in the skin of infested guinea pigs, Allen, Khalil and Wikel (1979) showed by indirect immunofluorescence that tick antigens were trapped by dendritic cells resembling Langerhans cells in the epidermis. Dendritic cells bearing antigens were also demonstrable in the dermis and draining lymph nodes. Epidermal Langerhans cell populations were monitored in the epidermis of tick infested guinea pigs by Nithiuthai and Allen (1984a). Very marked reductions in numbers of these cells occurred in sites surrounding tick attachments during primary infestations, and increased concentrations of cells were demonstrable early in secondary infestations of resistant animals. Such changes in Langerhans cell populations were similar to those shown by Silberberg, Baer and Rosenthal (1974) and Silberberg-Sinakin, Baer and Thorbecke (1978) to occur in guinea pigs subjected to primary and secondary applications of the contact allergen, dinitrochlorobenzene. In 1973, Schleger and Bean recorded that dendritic alkaline phosphatase-positive cells in the epidermis of cattle were markedly reduced in number following infestations with *B. microplus* ticks. These authors suggested that these alkaline phosphatase-positive cells were Langerhans cells, and recently this has been confirmed (Khalil, Nithiuthai & Allen 1982).

Nithiuthai and Allen (1984b, 1984c) showed that ultraviolet irradiation of guinea pig ear skin was associated with marked reductions in the numbers of adenosine triphosphatase-positive epidermal Langerhans cells. The populations of these cells were significantly reduced for at least 6 days following irradiation, and minimal inflammatory responses in the epidermis were evident. When primary infestations with *D. andersoni* larvae were confined to irradiated skin, the acquisition of resistance was significantly reduced. Also, the expression of tick resistance was significantly reduced when ticks were confined to skin which had been irradiated. Furthermore, in *in vitro* blastogenesis experiments, epidermal cell populations containing Langerhans cells have been shown to be capable of presenting tick salivary antigens to syngeneic lymphocytes from tick resistant animals. Epidermal cell populations depleted of Langerhans cells were unable to do this (Nithiuthai & Allen 1985). From these bits of information, it seems likely that Langerhans cells play at least some roles in the tick resistance response of guinea pigs. They may be of importance in presenting tick antigens in the afferent arm of the response, and may possibly act as antigen-laden target cells which provide early foci for immunological reactions in the epidermis and subsequent basophil reactions in the skin of resistant animals. Further studies of this subject in cattle which acquire high or low levels of tick resistance may be of interest.

Reactions to Bites of Mosquitoes and Ixodid Ticks

A speculative summary

Repeated injections of small doses of antigens into the skin induce Jones-Mote, or cutaneous basophil hypersenitivity reactions. Repeated mosquito bites also induce such reactions. Antigens in the mosquito saliva in the sensitized host cause local skin reactions, and sometimes serious systemic reactions. The mosquito, which completes its bloodmeal within a few minutes, leaves unharmed by the reaction. In contrast, Ixodid ticks normally remain attached to the host for 5–15 days before completing their bloodmeal. Hypersensitivity reactions to tick salivary antigens in the skin, involving the local release of histamine and other mediators from degranulating basophils or mast cells, appear to disrupt the feeding behaviour of attached ticks and represent one important mechanism of the tick resistance acquired by cattle and guinea pigs.

It is possible to induce hypersensitivity to salivary antigens of mosquitoes and ticks artificially, mimicking the naturally acquired hypersensitivities. It is possible also to immunize hosts artificially with antigens from the midgut or other organs of mosquitoes and ticks. Such antigens presumably are not introduced into the host during the normal bloodfeeding activities of the parasites but antibodies specific for these antigens, when ingested by the parasite, are believed to cause deleterious effects in the mosquito or tick which feeds on the immunized host. This immunization procedure may well be of practical value in the control of cattle ticks.

Finally, it seems that hosts which are sensitized to salivary antigens of mosquitoes may be protected to some extent from infections with mosquito-borne microorganisms. There is evidence that this may be so in the case of mosquito-borne Sindbis virus and *Plasmodium* infections. A suggested explanation for this phenomenon is that viruses or plasmodial sporozoites, developing within the salivary gland of the mosquito, incorporate mosquito antigens in their envelopes or membranes and that these antigens represent targets for specific humoral or cellular reactions in the mosquito-sensitized host.

There is evidence to suggest that tick-sensitized (tick-resistant) hosts may also be significantly resistant to infection with tick-borne microorganisms. Rabbits which had acquired tick resistance following infestations with pathogen-free *D. andersoni* were found to be significantly protected against *Francisella tularensis* infections when challenged with infected ticks (Bell, Stewart & Wikel 1979; Wikel 1980). Also, Mishaeva, Votyakov and Tarasenko (1981) have shown that guinea pigs sensitized to salivary antigens of pathogen-free *D. andersoni* were significantly protected from infections with tick-borne encephalitis virus following challenge infestations with virus-infected ticks. It may be, as these authors suggest, that the protection from infection was the result of disrupted feeding behaviour caused by the tick-resistance mechanism. Alternatively, drawing a parallel with the mosquito story, tick salivary antigens included in the virion envelope of transmitted viruses might act as targets for immunological reactions specific for these antigens.

It is possible that artifical immunization of cattle with tick antigens could result in significant tick-resistance together with reduction in transmission of tick-borne microorganisms. It would be interesting to investigate further the transmission of tick-borne pathogens to tick resistant animals. Possibly, reduced transmission of such organisms as *Theileria, Babesia* and *Rickettsia* would occur due to the cessation of salivation or detachment of ticks caused by the tick resistance mechanism. On the other hand, the incorporation or close association of tick salivary antigens with the limiting membranes of the microorganisms might represent suitable targets for the tick-resistant host's immune responses.

References

Ackerman, S., Clare, F.B., McGill, T.W. & Sonenshine, D.E. 1981. Passage of host serum components, including antibody, across the digestive tract of *Dermacentor variabilis* (Say). *Journal of Parasitology* **67**:737–40.

Alger, N.E. and Cabrera, E.J. 1972. An increase in death rate of *Anopheles stephensi* fed on rabbits immunized with mosquito antigen. *Journal of Economic Entomology* **65**:165–68.

Alger, N.E., Harant, J.A., Willis, L.C. and Jorgensen, G.M. 1972. Sporozoite and normal salivary gland induced immunity in malaria. *Nature* **238**:341.

Allen, J.R. 1966. Passive transfer between experimental animals of hypersensitivity to *Aedes aegypti* bites. *Experimental Parasitology* **19**:132–37.

Allen, J.R. 1973. Tick resistance: Basophils in skin reactions of resistant guinea pigs. *International Journal for Parasitology* **3**:195–200.

Allen, J.R. and West, A.S. 1966. Some properties of oral secretion from *Aedes aegypti* (L.). *Experimental Parasitology* **19**:124–31.

Allen, J.R. Doube, B.M. and Kemp, D.H. 1977. Histology of bovine skin reactions to *Ixodes holocyclus* Neumann. *Canadian Journal of Comparative Medicine* **41**:26–35.

Allen, J.R. and Humphreys, S.J. 1979. Immunisation of guinea pigs and cattle against ticks. *Nature* **280**:491–93.

Allen, J.R., Khalil, H.M. & Wikel, S.K. 1979. Langerhans cells trap tick salivary gland antigens in tick-resistant guinea pigs. *Journal of Immunology* **122**:563–65.

Allen, J.R. and Kemp, D.H. 1982. Observations on the behaviour of *Dermacentor andersoni* larvae infesting normal and tick resistant guinea pigs. *Parasitology* **84**:195–204.

Askenase, P.W. 1980. Immunopathology of parasitic diseases: Involvement of basophils and mast cells. *Springer Seminars in Immunopathology* **2**:417–22.

Askenase, P.W., Bagnall, B.G. and Worms, M.J. 1982. Cutaneous basophil-associated resistance to ectoparasites (ticks). I. Transfer with immune serum or immune cells. *Immunology* **45**:501–11.

Bell, J.F., Stewart, S.J. and Wikel, S.K. 1979. Resistance to tick-borne *Francisella tularensis* by tick-sensitized rabbits: Allergic Klendusity. *American Journal of Tropical Medicine and Hygiene* **28**:876–80.

Brossard, M. 1976. Relations immunologiques entre bovines et tiques, plus particulierement entre bovines et *Boophilus microplus. Acta Tropica* **33**:15–36.

Brown, A., Griffitts, T.H.D., Erwin, S. & Dyrenforth, L.Y. 1938. Arthus phenomenon from mosquito bites. *Southern Medical Journal* **31**:590–95.

Brown, S.J. & Askenase, P.W. 1981. Cutaneous basophil responses and immune resistance of guinea pigs to ticks: Passive transfer with peritoneal exudate cells or serum. *Journal of Immunology* **127**:2163–67.

Brown, S.J., Galli, S.J., Gleich, G.J. and Askenase, P.W. 1982. Ablation of immunity to *Amblyomma americanum* by anti-basophil serum: Cooperation of basophils and eosinophils in expression of immunity to ectoparasites (ticks) in guinea pigs. *Journal of Immunology* **129**:790–96.

Feinsod, F.M., Spielman, A. and Waner, J.L. 1975. Neutralization of togavirus by antivector antisera. *Annals of the New York Academy of Sciences* **266**:251–54.

French, F.E. 1972. *Aedes aegypti*: Histopathology of immediate skin reactions of hypersensitive guinea pigs resulting from bites. *Experimental Parasitology* **32**:175–80.

Friedmann, P.S. 1981. The immunobiology of Langerhans cells. *Immunology Today*. **2**:124–28.

Gordon, R.M. and Lumsden, W.H.R. 1939. A study of the behaviour of the mouthparts of mosquitoes when taking up blood from living tissue; together with some observations on the ingestion of microfilariae. *Annals of Tropical Medicine and Parasitology* **33**:259–78.

Gregson, J.D. 1967. Observations on the movement of fluids in the vicinity of the mouthparts of naturally feeding *Dermacentor andersoni* Stiles. *Parasitology* **57**:1–8.

Heilesen, B. 1949. Studies on mosquito bites. *Acta Allergologica* **2**:245–67.

Heller-Haupt, A., Varma, M.R.G. and Langi, A.O. 1981. Acquired resistance to Ixodid ticks in laboratory animals. *Transactions of the Royal Society of Tropical Medicine and Hygiene* **75**:147–48.

Hewetson, R.W. 1971. Resistance by cattle to cattle tick *Boophilus microplus*. III. The development of resistance to experimental infestations by purebred Sahiwal and Australian Illawarra Shorthorn cattle. *Australian Journal of Agricultural Research* **22**:331–42.

Hudson, A., Bowman, L. and Orr, C.W.M. 1960. Effects of absence of saliva on bloodfeeding by mosquitoes. *Science* **131**:1730–31.

Kemp, D.H., Koudstaal, D., Roberts, J.A. and Kerr, J.D. 1976. *Boophilus microplus*: The effect of host resistance on larval attachments and growth. *Parasitology* **73**:123–136.

Kemp, D.H. and Bourne, A.S. 1980. *Boophilus microplus*: The effect of histamine on the attachment of cattle tick larvae-studies *in vivo* and *in vitro*. *Parasitology* **80**:487–96.

Khalil, H., Nithiuthai, S. and Allen, J.R. 1982. Alkaline phosphatase-positive Langerhans cells in the epidermis of cattle. *Journal of Investigative Dermatology* **79**:47–51.

Koudstaal, D., Kemp, D.H. and Kerr, J.D. 1978. *Boophilus microplus*: Rejection of larvae from British breed cattle. *Parasitology* **76**:379–86.

Krinsky, W.L., Brown, S.J. and Askenase, P.W. 1982. *Ixodes dammini*: Induced skin lesions in guinea pigs and rabbits compared to erythema chronicum migrans in patients with Lyme Arthritis. *Experimental Parasitology* **53**:381–95.

McKiel, J.A. 1959. Sensitization to mosquito bites. *Canadian Journal of Zoology* **37**:341–51.

McKiel, J.A. and West, A.S. 1961. Nature and causation of insect-bite reactions. *Pediatric Clinics of North America* **8**:795–816.

Mellanby, K. 1946. Man's reaction to mosquito bites. *Nature* **158**:554, 751.

Mishaeva, N.P., Votyakov, V.I. and Tarasenko, A.B. 1981. [Transfer of resistance against *Ixodoidea* to vertebrates with serum and lymphocytes obtained from immune animals] (in Russian). *Zhurnal Mikrobiologii, Epidemiologii i immunobiologii* **3**:35–39.

Moorhouse, D.E. and Tatchell, R.J. 1969. Histological responses of cattle and other ruminants to the recent attachment of Ixodid larvae. *Journal of Medical Entomology* **6**:419–22.

Mote, J.R. and Jones, T.D. 1936. The development of foreign protein sensitization in human beings. *Journal of Immunology* **30**:149–67.

Newsome, W.H., Jones, J.K.N., French, F.E. and West, A.S. 1969. The isolation and properties of the skin-reactive substance in *Aedes aegypti* oral secretion. *Canadian Journal of Biochemistry* **47**:1129–36.

Nithiuthai, S. and Allen, J.R. 1984. Significant changes in epidermal Langerhans cells of guinea pigs infested with ticks (*Dermacentor andersoni*). *Immunology* **51**:133–41.

Nithiuthai, S. and Allen, J.R. 1984b. Effects of ultraviolet irradiation on epidermal Langerhans cells in guinea pigs. *Immunology* **51**:143–51.

Nithiuthai, S. and Allen, J.R. 1984c. Effects of ultraviolet irradiation on the acquisition and expression of tick resistance in guinea pigs. *Immunology* **51**:153–59.

Nithiuthai, S. and Allen, J.R. 1985. Langerhans cells present tick antigens to lymph node cells from tick sensitized guinea pigs. *Immunology* **55**:157–63.

Paine, S.H., Kemp, D.H. and Allen, J.R. 1983. *In vitro* feeding of *Dermacentor andersoni* (Stiles): Effect of histamine and other mediators. *Parasitology* **86**:419–28.

Richerson, H.B., Dvorak, H.F. and Leskowitz, S. 1969. Cutaneous basophil hypersensitivity. I. A new look at the Jones-Mote reaction, general characteristics. *Journal of Experimental Medicine* **132**:546–57.

Riek, R.F. 1962. Studies on the reactions of animals to infestation with ticks. VI. Resistance of cattle to infestation with the tick, *Boophilus microplus*. (Canestrini) *Australian Journal of Agricultural Research* **13**:532–50.

Roberts, J.A. 1968. Acquisition by the host of resistance to the cattle tick, *Boophilus microplus* (Canestrini). *Journal of Parasitology* **54**:657–62.

Roberts, J.A. and Kerr, J.D. 1976. *Boophilus microplus*: Passive transfer of resistance in cattle. *Journal of Parasitology* **62**:485–89.

Rockwell, E.M. and Johnson, P. 1952. The insect bite reaction. II. Evaluation of the allergic reaction. *Journal of Investigative Dermatology* **19**:137–55.

Schleger, A.V. and Bean, K.G. 1973. The melanocyte system of cattle skin. I. Amelanotic dendritic cells of epidermis. *Australian Journal of Biological Sciences* **26**:973–83.

Schleger, A.V., Lincoln, D.T., McKenna, R.V., Kemp, D.H. and Roberts, J.A. 1976. *Boophilus microplus*: Cellular responses to larval attachment and their relationship to tick resistance. *Australian Journal of Biological Sciences* **29**:499–512.

Silberberg, I., Baer, R.L. and Rosenthal, S.A. 1974. The role of Langerhans cells in contact allergy. I. An ultrastructural study in actively induced contact dermatitis in guinea pigs. *Acta Dermatovenereology (Stockholm)* **54**:321–31.

Silberberg-Sinakin, I., Fedorko, M.E., Baer, R.L., Rosenthal, S.A., Berezowsky, V. and Thorbecke, G.J. 1977. Langerhans cells: Target cells in immune complex reactions. *Cellular Immunology* **32**:400–416.

Silberberg-Sinakin, I., Baer, R.L. and Thorbecke, G.J. 1978. Langerhans cells: A review of their nature with emphasis on their immunological functions. *Progress in Allergy* **24**:268–94.

Stingl, G. 1980. New aspects of Langerhans' cell function. *International Journal of Dermatology* **19**:189–213.

Sutherland, G.B. and Ewen, A.B. 1974. Fecundity decrease in mosquitoes ingesting blood from specifically sensitized mammals. *Journal of Insect Physiology* **20**:655–60.

Suzuki, S., Negishi, K., Tomizawa, S., Shibasaki, M., Kuroume, T. and Matsumura, T. 1976. A case of mosquito allergy: Immunological studies. *Acta Allergologica* **31**:428–41.

Tager, A., Lass N., Gold, D. and Lengy, J. 1969. Studies on *Culex pipiens molestus* in Israel. IV. Desensitization attempts on children showing Strophulus-like skin eruptions following bites of the mosquito. *International Archives of Allergy* **36**:408–14.

Tatchell, R.J. and Moorhouse, D.E. 1968. The feeding processes of the cattle tick *Boophilus microplus* (Canestrini). Part II. The sequence of host-tissue changes. *Parasitology* **58**:441–59.

Theis, J.H. and Budwiser, P.D. 1974. *Rhipicephalus sanguineus*: Sequential histopathology at the host-arthropod interface. *Experimental Parasitology* **36**:77–105.

Trager, W. 1939. Acquired immunity to ticks. *Journal of Parasitology* **25**:57–81.

Utech, K.B.W., Seifert, G.W. and Wharton, R.H. 1978. Breeding Australian Illawarra Shorthorn cattle for resistance to *Boophilus microplus*. I. Factors affecting resistance. *Australian Journal of Agricultural Research* **29**:411–22.

Wagland, B.M. 1975. Host resistance to cattle tick (*Boophilus microplus*) in Brahman (*Bos indicus*) cattle. I. Responses of previously unexposed cattle to four infestations with 20,000 larvae. *Australian Journal of Agricultural Research* **26**:1073–80.

Wikel, S.K. 1980. Host resistance to tick-borne pathogens by virtue of resistance to tick infestation. *Annals of Tropical Medicine and Parasitology* **74**:103–4.

Wikel, S.K. 1981a. The induction of host resistance to tick infestation with a salivary gland antigen. *American Journal of Tropical Medicine and Hygiene* **30**:284–88.

Wikel, S.K. 1981b. Histamine content of tick attachment sites and the effects of H_1 and H_2 histamine antagonists on the expression of resistance. *Annals of Tropical Medicine and Parasitology* **76**:179–85.

Wikel, S.K. 1982. Immune responses to Arthropods and their products. *Annual Review of Entomology* **27**:21–48.

Wikel, S.K. and Allen, J.R. 1976a. Acquired resistance to ticks. II. Effects of Cyclophosphamide on resistance. *Immunology* **30**:479–84.

Wikel, S.K. and Allen, J.R. 1976b. Acquired resistance to ticks. I. Passive transfer of resistance. *Immunology* **30**:311–16.

Wikel, S.K., Graham, J.E. and Allen, J.R. 1978. Acquired resistance to ticks. IV. Skin reactivity and *in vitro* lymphocyte responsiveness to salivary gland antigen. *Immunology* **34**:257–63.

Willadsen, P. 1980. Immunity to ticks. *Advances in Parasitology* **18**: 293–313.

Willadsen, P., Williams, P.G., Roberts, J.A. and Kerr, J.D. 1978. Responses of cattle to allergens from *Boophilus microplus*. *International Journal for Parasitology* **8**:89–95.

Willadsen, P., Wood, G.M. and Riding, G.A. 1979. The relation between skin histamine concentration, histamine sensitivity and the resistance of cattle to the tick, *Boophilus microplus*. *Zeitschrift für Parasitenkunde* **59**:87–93.

7 Histopathology of Benign Tumors Due to *Demodex antechini* in *Antechinus stuartii*

Wm.B. Nutting

Introduction

Since 1958, impetus to studies on demodicosis in Australian mammals has been provided the author by Dr John F.A. Sprent and his parasitology group at the University of Queensland. The substance of this report on benign tumors produced by *Demodex antechini* (Nutting & Sweatman 1970) in the skin of Stuart's brown marsupial mouse (*Antechinus stuartii*) is based on specimens obtained through contacts (see Acknowledgements) made during three sabbatical visits for research under the sponsorship of Dr Sprent's department.

Topics covered below include new information on mite and bacterial impacts, host cellular responses, host tissue changes, and host-parasite interactions. A brief synoptic summary-discussion places these results in a sequential pattern of initial mite invasion and micropathogenesis, through tumor origin, development, degradation, and skin plaque formation.

Materials and Methods

Several hundred slides of unparasitized and parasitized skin from seven body areas (dorsal, ventral, genital, pouch margin, eyelid, muzzle and pinna) and lesions of five body areas (pinna base, dorsal, ventral, genital and pouch margin) were studied. Skin samples were fixed (70% ethyl alcohol or 4% formaldehyde), blocked in paraffin, sectioned at 4–12 μm, and variously stained (as Nutting & Beerman 1965; and others, see Results) using standard methods (Humason 1962). Some sections of each area or lesion were stained with conventionl hematoxylin and eosin. Slides were studied with light, phase, and interference microscopy.

All reported results are based on measurements or observations of more than 21 sections each of mites, cells, tissues, or sequential histopathological changes. Because most of these assessments are based on continua, they are noted as ranges of values rather than average values.

All skin samples were necropsied and preserved as samples selected from 160 wild caught specimens from New South Wales, Australia, obtained and/or maintained in the Zoology Department Laboratories at the Australian National University (see Nutting & Woolley 1965) under the supervision of Dr Patricia Woolley.

Macro-lesions and clinically normal skin samples were necropsied from 13 (5 male, 8 female) laboratory maintained specimens. Because synhospitaly is characteristic of this mite genus (Nutting 1979), *Demodex antechini* was determined to species using characters in Nutting and Sweatman (1970) for each sample used in this study.

Results

A. Impact of mites or associated bacteria

In all instances, and for any and all seven pilosebaceous areas examined, only adult mites are found entering the pilosebaceous orifice (fig.7.1). In initial invasions of non-infested follicles, a single mite cuts body-size (diameter av. 25 μm) pits directly adjacent and parallel to the hair (diameter 20–25 μm) and epithelia (at the surface, gross diameter av. 125 μm) apparently feeding on the cytoplasm of the cells. Subsequent adult invasions extend the follicular cavity, in ratio to their numbers, decimating cells distal to the sebaceous gland duct and paralleling not penetrating the stratum germinativum. Eggs laid in these distal luminae hatch to larvae which penetrate epithelial or gland cells proximal to their site of hatching. These larvae penetrate cells with their long axis nearly perpendicular to the germinativum, harvesting successive cells and moulting to nymphs and then adults *in situ*. Adults either deposit ova in the epithelial pits, or pilosebaceous luminae and then (presumably) move out of the follicle to relocate. Up to this point mite impact has provided cell damage, moderate keratinization (claw abrasion), some minor impedence of sebum flow, and widening of the pilosebaceous orifice.

Secondary bacterial infections are found occasionally in these early invasions sharing single pilosebaceous systems with the mites. They produce rapid destruction of distal epithelia, and dermal elements. Cells upon which the mites are feeding are destroyed, the dermis shows cellular elements of a typical (bacterial) inflammatory response, and the follicular canal becomes plugged with pus. No bacteria were found in either (1) sections of mature tumors or (2) in the absence of *Demodex antechini* in the pilosebaceous complex. Bacteria were Gram positive.

The suppurative plugging due to bacterial activity may either kill the mites (internal tissue disintegrated) or in sections of follicles, distal to mite infestations, macrophage clearing and fibroblast activity may trap live mites below the tissue repair. These trapped mites, as well as first generation adults from multiple early invaders, which continue to reproduce within their home follicle, provide bacteria-free rapidly increasing mite populations subsisting on the renewing cell populations of the strata germinativa.

In non-infected but heavily mite infested follicles, the first generation of adult mites often seem incapable of exodus before egg deposition. Their

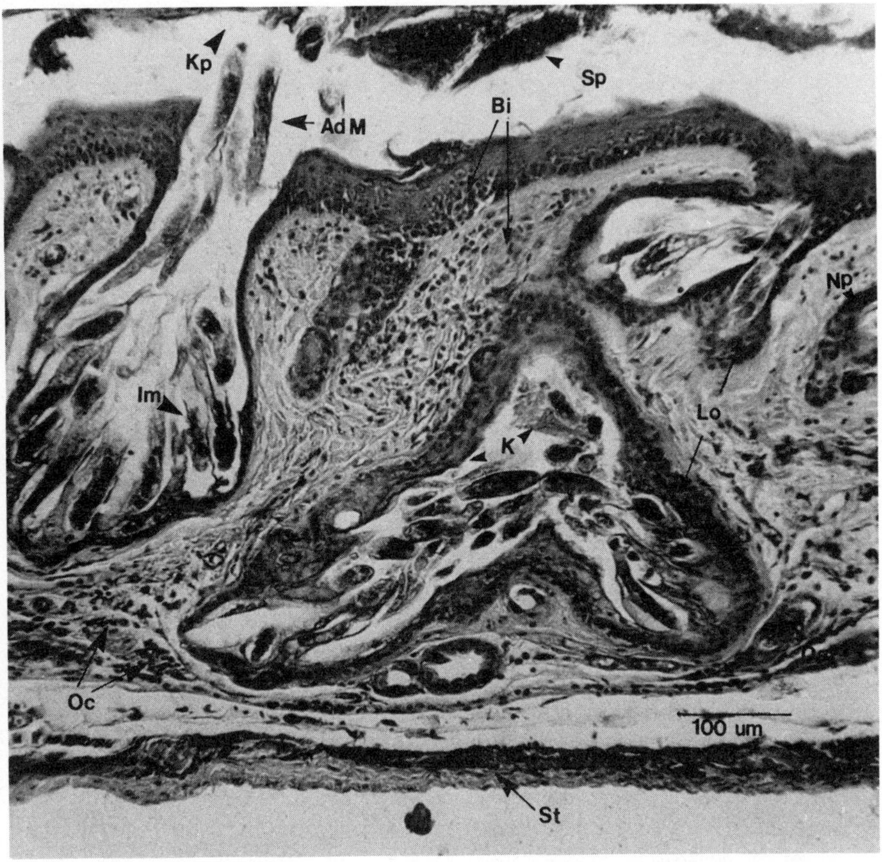

Figure 7.1 Photomicrograph of a skin section showing incipient demodectic tumors. Adult mites (*AdM*) in distal follicles and immatures (*Im*) more proximal; keratin (*Kp*) and suppurative (*Sp*) debris plugs; reaction to bacteria (*Bi*); epithelial lobules (*Lo*) with keratinized (*K*) epithelium plus debris and ova in follicular cavities; arterial stasis (*Oa*) and reaction to vascular occlusion (*Oc*); subdermal tissue (*St*) including muscularis, fascia, and adipose tissue. Area near *Np* (pilosebaceous system with hair, epithelial and sebaceous cells) is normal for the non-infested skin complex. (Haematoxylin-Eosin stain)

activity, producing keratinization, cellular debris from the feeding immatures and body pressures of both, apparently further restrict the pilosebaceous exit ways. Feeding activity and body pressures of the mites evict the hairs, produce lobulation of both follicular and sebaceous epithelium, stimulate the germinatival epithelia, and occasion the newly discovered mite-associated light cells (Nutting, 1985).

The impacts of bacteria in association with the mites seems limited primarily to early subclinical demodicosis usually prior to demodectic lobulation. Impact of the rapidly increasing population of the trapped *Demodex antechini* continues at an accelerating rate (a nodule may form in about 30 days — see Nutting & Woolley 1965) with the following histopathological changes:

1. *Cell harvesting*, primarily by immature mites, keratinization by adult claws, body pressures of both continue and form the major stimuli to lobulation. As lobulation proceeds a variety of host cellular responses occur (see below) with lobules fanning down into the subdermal fascia thus pressing and stretching the overlying skin producing drastic host tissue changes (further below). Clinically the lesion is then a hard core, hair-depleted nodule.
2. *Depletion of lobular epithelium* by the mites leads to local rupture or mite cheliceral penetration into the dermis or the vascular system leading to marked host cellular responses, a clinically "papular" condition, and finally recession of the demodectic tumor to a skin plaque.

The changes occuring in the extra lobular tissue seem produced exclusively by physical pressures of mite activity and host cellular responses. These changes include minor occlusion of capillaries (fig.7.1), interdigitation of new and often sinus-like vascular vessels, and of connective tissue (fig.7.4). Upon lobular rupture individual mites are invaded by macrophages (see Nutting 1975, for *Demodex caprae*) and the resultant raw exoskeletons then stimulate a foreign body reaction. No marked host immune responses to the mites, as also noted by Nutting and Sweatman (1970), was evident even in mice with multiple demodectic infestations.

Upon papular recession only moribund adult mites are found in small, dermal pits, the walls of which are under repair by fibroblasts and small macrophages.

B. Host cellular responses

The cells of the pilosebaceous germinativum provide a "renewable resource" for demodecid feeding and development, for tissue repair after mites feed, and between adjacent feeding mites (fig. 7.1) in forming the walls of each developing lobule. Most spectacular of all, the cells termed epidermal giant cells (Nutting & Beerman 1965) are formed from daughter cells of this germinatival layer. These giant cells are large (80–200 μm), multinucleate (8–35 nuclei) and apparently potentially phagocytic (fig.7.2). We now feel assured that they form in either of two ways: (1) by penetrating through several epithelial cell layers or (2) by penetrating a syncytial mass adjacent to the basement membrane. Sections reveal that such giant cells form in lobulations of both sebaceous and follicular epithelia. In either, even cell penetration seems to take place only at the mite mouthparts with the giant cell membrane pulled down or reconstituted along the sides of the mite. In a few cases of multiple penetration each mite is templated by a separate cell membrane. Nuclei

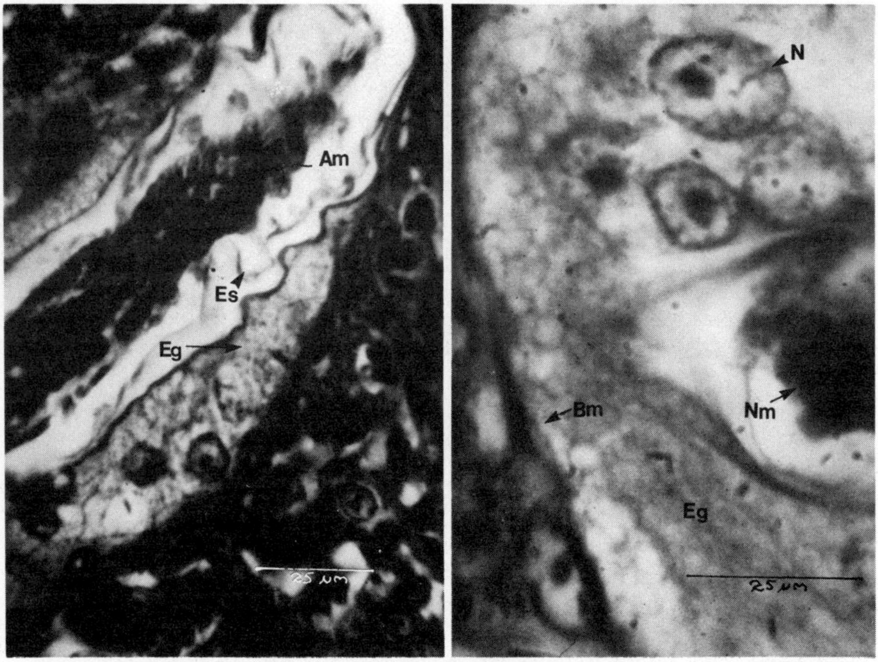

Figure 7.2 Photomicrographs of sections of ectodermal giant cells (*Eg*). **7.2a** An adult mite (*Am*) maturing within the giant cell surrounded by ecdysed larval and nymphal exoskeltons (*Es*). Note templating of investing cell membrane. (Masson stain) **7.2b** The base, against basement membrane (*Bm*). Nuclei (*N*) in all stages of distortion. The gnathosoma of a nymphal stage mite (*Nm*) is adjacent to area of cell membrane penetration. Note vacuolated cytoplasm of giant cell. (Hematoxylin-Eosin stain)

nearest mite mouthparts are rounded, enlarged and disintegrated even sooner than those bypassed in mite penetration (fig. 7.2). The cytoplasm is also most vacuolated nearest the mouthparts. Depletion by mite ingestion apprently occurs at this point aided by mite salivary products. Mites mature (larvae ecdyse to nymphs and these to adults) at the expense of these giant cells (fig.7.2a).

The triangular cells reported by Nutting and Sweatman (1970) as lying in pockets between cells of the lobular germinativum cells match in morphology (eccentric nuclei) and staining (Giemsa) characteristics, the small macrophages of the interlobular dermis especially that of degenerating dermal tissue. They seem to fill intercellular spaces in small irregularly distributed groups possibly in response to physical rather than biochemical stimuli of either host epithelial cells or invading mites.

These same small macrophages are common in both the interlobular and supralobular connective tissue. Such dermal areas and even those of the reconstituting skin plaque (below) are peppered also with histiocytes

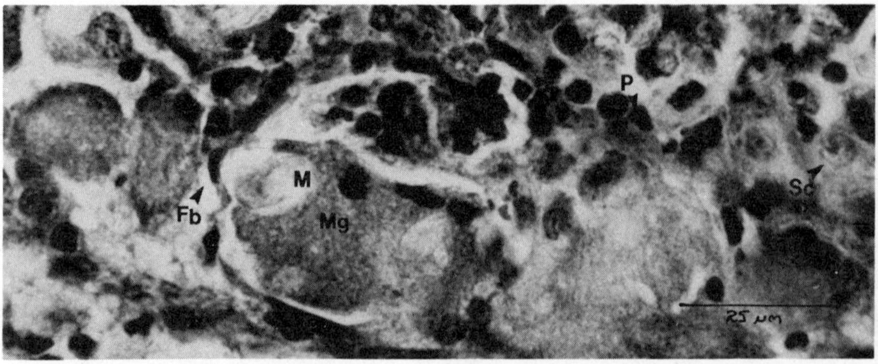

Figure 7.3 Photomicrograph of section showing granulomatous tissue with (*Mg*) a mesodermal giant cell containing mites (*M*). Many macrophages (as *P*), and fibroblasts (as *Fb*); and a few epithelial (sebaceous) cells (as *Sc*) are also shown. Note the dense-staining nuclei which differ from mesodermal giant cells in eutherian mammals. The homogeneous granular cytoplasm also differentiates these from ectodermal giant cells. (Hematoxylin-Eosin stain)

(centric nuclei) and fibroblasts (elongate nuclei). The latter form the connective tissue fibers. Upon investment of the connective tissue, macrophages invade and clear the internal tissues of the mites.

Figure 7.3 shows the typical foreign body reaction which seems in *Anthechinus* to occur by macrophage fusion to become a mesodermal, phagocytic, multi-nucleate giant cell. These phagocytose mites, moulted exoskeletons, and larger particles of mite-produced keratin. These cells (1) measure 40–160 μm, (2) possess 6–15 nuclei, (3) have granular, homogenous eosinophilic cytoplasm, and (4) show densely staining nuclei similar to those of macrophages.

Host cells in the mite-associated bacterial inflammatory response were typical of those usually found associated only with non-metazoan parasitic skin involvement (i.e. few granulocytic polymorphs). This same picture pertained to the mite-caused pseudoinflammatory response occasioned by blood-sinus rupture. This also seemed low in polymorphonuclear granulocytes (eosinophils?).

C. Host tissue changes

The sequence of host tissue changes from normal non-infested skin to plaque formation mirrors a continuous pattern of aggregation of individual cellular responses to mite activity, clearly distinguishable and mensurable as distinctive tissues only at the nodular, papular, and plaque stages. Near-normal tissue is illustrated to the right in figure 7.1 with a few measurements of normal skin components ascertainable by scale. Comparative measurements for nodular, papular, and plaque tissues also can be estimated using the scales in figures 7.4 and 7.5.

1. Nodular tissues (fig. 7.4)

Incipient nodules show only spotty, local, and often transient tissue changes, but can lead in less than one month (Nutting & Woolley 1965) to a hard core clinically recognizable nodular lesion. Sections show that as compared to normal skin the epidermis is (1) thinned with fewer viable cell layers now more squamous than cuboidal, (2) infiltrated to a minor degree by macrophages, and (3) with the stratum granulosum overlain by

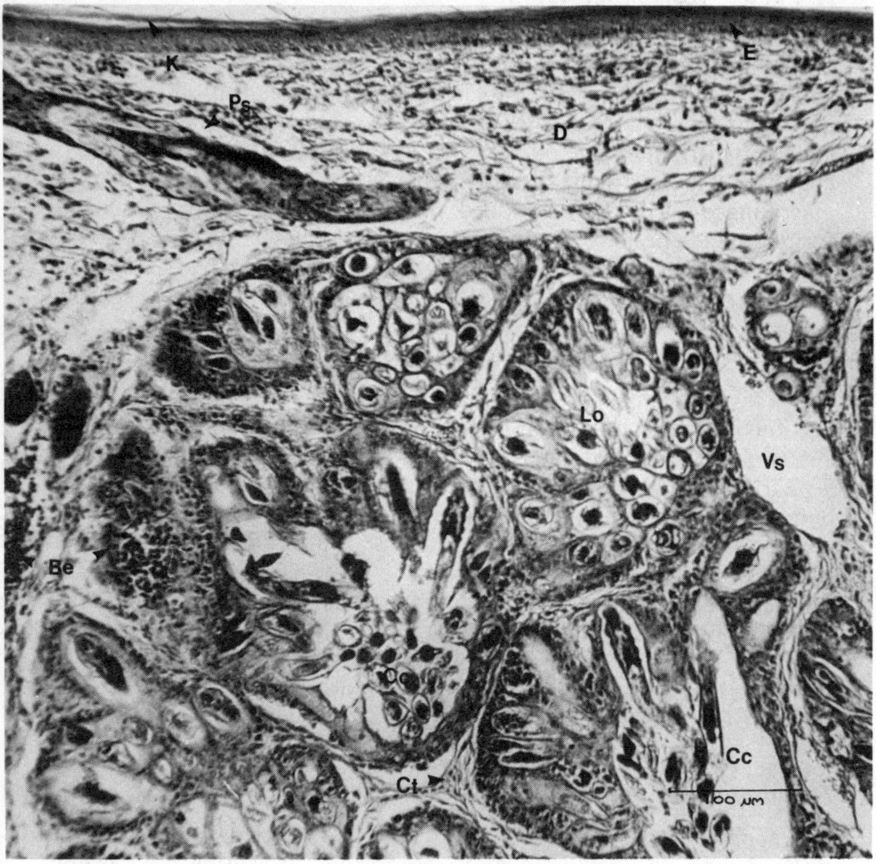

Figure 7.4 Photomicrograph of a section taken at the edge of the (clinically) protruding nodule (nodule central maximum right) showing epidermis (*E*), keratin (*K*) and disintegrating dermis (*D*); occlusive (vascular) derangement or mite penetration with mixture of blood elements and macrophages at *Be*; central cavities (*Cc*) with mites, wisps of exoskeleton and keratin. Surrounding epithelium is thinned, squamous, and keratinized. Epithelial lobule (*Lo*), interlobular connective tissue (*Ct*) and interlobular blood vessels, including large sinuses at *Vs*. Note immature mites in lobules and adults and ova usually in lacunae (*Cc*). The pilosebaceous system (*Ps*) is laterally compressed. (Hematoxylin-Eosin stain)

denser strap-like layered keratin. The interfollicular distances are five or more times normal in extent with pilosebaceous components compressed. Many eccrine glands, dermal capillaries, and dermal and subdermal tissues are destroyed possibly by capillary occlusion leading to cell death and macrophage invasion. Fingers of nodular tissue are 100–200 μm long by 45–270 μm in diameter in basal section. Between nodules lie thin strips of connective tissue surrounding capillaries and large vascular sinuses.

2. *Papular tissues* (fig. 7.5)

In the late stages of nodular disintegration the large central internal cavity of the papule holds tissue debris, including pieces of moulted mite exoskeletons and keratin, live mites and free blood elements. Macrophages are concentrated around the cellular debris, and mites on the cavity periphery. Mesodermal giant cells are found forming near or engulfing these stranded mites and particulate debris. Above the macrophage-filled area an added modicum of connective tissue has formed and especially around enlarged arterioles and venules whose contents seem in stasis. The epidermis has lost its surface keratin but now stains as though moribund. Macrophage activity is also more evident in this area.

3. *Plaque tissues* (fig. 7.5)

Sections through this area show some reconstitution of the lower epidermis and dermis. Pockets of granulomata and of degenerating-walled cavities containing moribund (only) adult mites are found in the dermis. Connective tissue fibers, fibroblasts, and macrophages are most numerous around these cavities, otherwise the dermis and subdermis is quite returned to normal.

One must realize in this brief account that the superior mid-central area of the lesion usually shows maximal evidences of host tissue changes. Figures 7.4 and 7.5 are made from sections at the lesion periphery to provide a best approximation of the total lesion histopathology in terms of host tissue transitions.

D. Host-parasite interactions

It is apparent from the above that even at the incipient tumor stage no single demodicid or its progeny is responsible for causing these benign tumors. *Demodex antechini* females, as with other members of the genus (Desch & Nutting 1977), produce few ova (about 20) per female; these are laid singly, and at intervals of at least several days. From observations of *Demodex brevis* (English & Nutting 1981) and hundreds of *D. antechinus* sections, the usual picture of an infestation presents as one or two adults with a maximum of one or two ova or one or two immatures, per pilosebaceous complex (fig. 7.5). With adults, only, in the distal

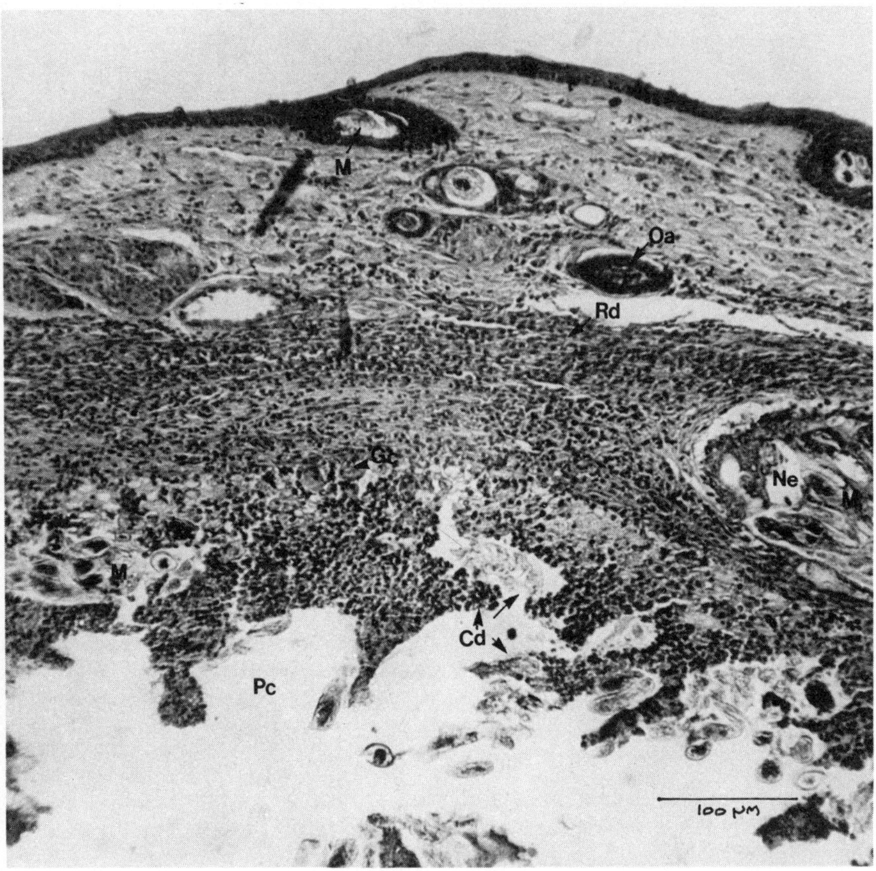

Figure 7.5 Photomicrograph of a section taken at the (clinical) boundary of the papule junction with non-papular skin (papule maximum left) showing mites (*M*), papule cavity (*Pc*) containing free cellular debris (*Cd*) including mite exoskeletons, live mites, vascular elements and macrophages. Mesodermal giant cells (*Gc*) and mites lie at the periphery entrapped by disintegrating (*Ne*) nodular epithelium (*right*) and macrophages. Mites as a semi-solitary infestation are shown above. The reconstituting dermis (*Rd*) is shown directly below an arteriole with vascular elements in stasis (*Oa*). Right half of this illustration is typical of the histo-pathology of the (clinical) demodectic skin plaque. (Hematoxylin-Eosin stain)

pilosebaceous canal (fig. 7.1) and some of these carrying well-developed ova, we believe that they normally move from the home pilosebacous system soon after maturation and invade nearby hair follicles.

At the individual level, also, minute size, weak chelicerae, lack of an anus, and small volume (only local and weak "toxicity") of salivary or body chemistry seems to be responsible for pathology only at the cellular level.

If, however, as noted above, multiple adult female mites invade and are trapped within a hair follicle the physical, chemical, and biotic host-parasite interactions produce the unusual histopathology reported herein. A few plausible, with some histopathologic evidence of, host-parasite interactions follows.

1. Host-mite physical interrelationships

Host tissues certainly respond both hyperplastically and mechanically to the pressures engendered by population increase of the incarcerated mites. The hyperplastic lobules press distally occluding circulatory supply to the overlying skin and proximally into the subdermal fascia, adipose tissue, and muscularis. The aggregate pressures of mite bodies providing nodule formation seems offset above by skin elasticity and below by the subdermal muscularis and deeper muscular body activity.

Of these shared interactions it seems likely that the physical characteristics of the local (areal) antechinus skin may be of prime importance in the extent and prevention of tumor formation. Maximal tumor formation was found in the relatively thin and flexible dorsal body skin, intermediate and moderate tumors in the dorsal, more resistant skin in the dorsal head region, and only incipient or secondary bacterial infections were found in the genito-cloacal pouch-marginal, or ventral skin. Minor semi-solitary infestations were common in all seven body areas sectioned including supra-lesional skin in all stages of tumor formation and degredation.

2. Host-mite physiology

The playback between mite and host physiology is obscured by the concomitant synergistic (below) and physical factors (above) involved in tumor formation. Despite this it is apparent that, in our sample, only those marsupial mice under incarceration developed tumors. Hormones, incriminated in demodicids of humans (Nutting & Green 1976) and corticosteroids (Hakugawa, — again in humans, 1978) seem to play only minor, if any, role in antechinal demodicosis. Five males and eight females developed tumors, and male *Antechinus* sp. show higher levels of systemic corticosteroids (according to Lee, Bradley & Braithwaite 1977).

It is plausible that the host species specific *Demodex antechini* has evolved under the selective pressures of rapid host maturation and short-term (2-year) life span. This rapid development, maturation and lifestyle may under environmental stress (as in captivity) biochemically speed up mite reproductive activity outrunning the normal host cellular defences.

3. Synergism

Measurements of the pilosebaceous orifices of incipient tumors (range 80–150 μm; fig.7.1) and mite-bacteria (synergistic) damaged follicles (range 40–100 μm) in contrast to orifices of the usual semisolitary mite infested follicles (>30 μm; fig. 7.5) leads us to the conclusion that a multiple mite infestation, seemingly necessary for tumor formation, is the result of developmental abnormality of the pilosebaceous apparatus and/or bacterial synergism. In theory, demodicid mites may transmit bacteria, and seed them (unwittingly) in the high-nutrient cell cytoplasm as the mites feed (Nutting 1976). Rapid cell destruction by both mites and bacteria could thus widen the entryway or even provide chemosensory cues to encourage multiple invasions of adult mites.

4. Immunity

Characteristics of these mites (above) coupled with the usual semi-solitary nature of their pilosebaceous invasion indicate, as we note from sections, no contact with the physiological systems or mechanisms usually associated with either inherited or acquired immunity.

Even under the conditions of synergism (above), penetration of these mites into the dermis, and even *en masse* directly adjacent to blood sinuses and/or capillaries, produced no perivascular or pronounced dermal infiltrations (also, Nutting & Sweatman 1970). This seems especially pertinent in the case of multiple-tumored mouse specimens. Sections of these fail to show any local or systemic cellular or humoral evidence of an immune response. We currently surmise that no specific to demodex immuno-deficient or immuno-protective mechanisms are operable in antechinal demodicoses.

Although the topics above hinge on speculation, some elements for each, and as stimulus for further research, is available both from *in litt.* and this report of antechinus lesions.

Discussion

The present study not only confirms the demodicid species, *Demodex antechini*, as causative organism for the origin and development of benign tumors in *Antechinus stuartii* but also adds, clarifies, or rectifies information in the general sequence proposed by Nutting and Beerman (1965) as follows.

1. Only adult mites invade the pilosebaceous system. These feed, and females deposit ova which hatch penetrating and destroying cells, as larvae, of the pilosebaceous epithelium. Upon *in situ* maturation of these, resulting adults move out to invade other pilosebaceous habitats.
2. Multiple adult invasions either trapped by cellular debris, mite bodies and/or products of secondary bacterial infections, continue to reproduce, their immatures subsisting on the contents of epithelial cells.

3. Incipient tumors (clinically obscure) are formed by sequential and multiple invasion of the epithelial cells by the immature mites. These produce host epithelial hyperplasia, with lobulations, interlobular dermal and vascular tissue, and ectodermal giant cells. These last are formed either by sequential cell penetration or penetration of the syncytial pilosebaceous epithelium.
4. Multiple epithelial lobulation with concurrent increase of interlobular connective and vascular tissue (the clinically "nodular" stage) lists, stretches and deforms the overlying skin. Nodule pressures occasion vascular occlusion producing deterioration of dermal and interlobular tissues leading to massive macrophage infiltration of dermal tissues.
5. Decimation of epithelial cells, dermal deterioration, and penetration of interlobular blood sinuses and capillaries engenders a pseudo-inflammatory response. Macrophages invade and clear tissue debris and internal tissues of those mites trapped in this clinically papular lesion. Adult mites and ova free in the papule cavity survive.
6. A foreign body reaction produces phagocytic mesodermal giant cells clearing mite exoskeletons and ova stranded in the inflammatory tissue and providing pockets of granulomatous tissue. Recession without rupture of the papule produces a thin epidermis and a dermis with small cavities of deteriorating epithelium enclosing moribund adult mites.
7. These skin plaques are slowly cleared of mites and the tissue reconstitutes, but they remain sparsely haired with a centrally thickened dermis.

As suggested by Nutting and Sweatman (1970) neither massive perivascular infiltration nor evidence of a humoral or systemic immune response to the mites could be found in our sections. The pseudo-inflammatory response seemed in all instances to be centered on disintegrating host cells or cell products rather than on the mites. Mites (either singly or grouped) stranded in the dermis were free of blood cell halos even in sections from mice with multiple lesions. Small numbers of macrophages entered and reduced mite tissues. Adult mites and ova free in the papule cavities remain viable even if surrounded by blood elements.

In conclusion this unique pattern of origin and degradation of demodectic tumors occurred only in the rapidly-maturing, short-lived, thin-skinned marsupial mice held several months post-maturity in captivity. Wild caught or captive immatures were clinically normal even though mites were retrieved from the former (Nutting & Woolley 1965). Because host species specific demodicids seem omnipresent in all mammals but monotremes, skin tumors on any other captive mammals with the characteristics of *Antechinus* may prove demodectic in origin, and be useful for comparative studies especially of the controversial mechanisms of immunity (see Scott, Farrow & Schultz 1974) as postulated for demodectic mange of dogs.

Acknowledgments

As indicated above, gratitude is deeply felt and here re-expressed for the friendship, help, concern, and stimulus to scientific research of Dr John F.A. Sprent over the past quarter century. Among most helpful contacts made through his good offices were: Dr Robert Domrow, Queensland Institute of Medical Research, who recognized the mites as demodecids and contacted the writer for collaboration with Dr Woolley; Dr Patricia Woolley, Monash University, who first located, preserved and sent the *Antechinus* tumors.

Prime movers in the present paper have been my wife, Margaret Nutting, for encouragement, technical help and preparations of illustrations, and Dr Clifford Desch for photography. To all above and including all faculty and staff since 1958 of Dr Sprent's Department of Parasitology, University of Queensland, my sincerest thanks and long-term gratitude.

References

Desch, C.E. and Nutting, W.B. 1977. Morphology and functional anatomy of *Demodex folliculorum* (Simon) of man. *Acarologia* **19**:422–62.

English, F.P. and Nutting, W.B. 1981. Demodicosis of ophthalmic concern. *American Journal of Ophthalmology* **91**:362–72.

Hakugawa, S. 1978. *Demodex folliculorum* infection on the face. Abnormal parasitism of *Demodex folliculorum* observable with persons using topical steroid preparation habitually on the face. *Western Japanese Dermatology* **40**:275–84.

Humason, G.L. 1967. *Animal tissue techniques*. 2nd ed. London: W.H. Freeman & Co.

Lee, A.K., Bradley, A.J. and Braithwaite, R.W. 1977. Corticosteroid levels and male mortality in *Antechinus stuartii*. In *The biology of marsupials*, ed. B. Stonehouse and D. Gilmore, 209–20. Baltimore: University Park Press.

Nutting, W.B. 1975. Pathogenesis associated with hair follicle mites (Acari:Demodicidae). *Acarologia* **17**:493–507.

Nutting, W.B. 1976. Hair follicle mites (Acari:Demodicidae) of man. *International Journal of Dermatology* **15**:79–98.

Nutting, W.B. 1979. Synhospitaly and speciation in the demodicidae (Trombidiformes). In *Proceedings 4th International Congress of Acarology*, ed. E. Piffl, 267–72. Budapest: Akademiai Kiado.

Nutting, W.B. 1985. Comparative topologic distribution and histopathology: Demodicidae. In *Acarology VI*, Vol. 2, ed. D.A. Griffiths and C.E. Bowman, Chichester, England: Ellis Horwood Ltd.

Nutting, W.B. and Beerman, H. 1965. Atypical giant cells in *Antechinus stuartii* due to demodicid mites. *Journal of Investigative Dermatology* **45**:504–9.

Nutting, W.B. and Green, A. 1976. Pathogenesis associated with hair follicle mites (*Demodex* spp.) in Australian Aborigines. *British Journal of Dermatology* **94**:307–12.

Nutting, W.B. and Sweatman, G.K. 1970 *Demodex antechini* sp. nov. (Acaris

Demodicidae) parasitic on *Antechinus stuartii* (Marsupialia). *Parasitology* **60**:425–29.

Nutting, W.B. and Woolley, P. 1965 Pathology in *Antechinus stuartii* (Marsupialia) due to *Demodex* sp. *Parasitology* **55**:383–89.

Scott, D.W., Farrow, B.R.H. and Schultz, R.D. 1974. Studies on the therapeutic and immunologic aspects of generalized demodectic mange in the dog. *Journal of the American Animal Hospital Association* **10**:233–44.

8 *Angiostrongylus cantonensis* in Rats: How Do Parasites Avoid Immunological Destruction?

W.K. Yong and C. Dobson

Introduction

Immunity to blood and tissue helminths has been comprehensively reviewed by different workers in the last decade (Dobson 1972; Olson & Izzat 1972; Gemmell & MacNamara 1972; Smithers & Terry 1976; Ogilvie & Rose 1978; Williams 1979; Soulsby, Monsell & Lloyd 1981; Dineen & Wagland 1982). These authors dealt mainly with the schistosomes, filarial worms, gastrointestinal nematodes and cestodes; immunity to *Angiostrongylus cantonensis*, a metastrongyle nematode parasite in the lungs of rats is rarely mentioned. Nevertheless knowledge on the immunity of rats after *A. cantonensis* infections has advanced significantly. Information is available about how acquired resistance is stimulated, and about the role of humoral and cell-mediated immunity in regulating *A. cantonensis* infection. It is perhaps pertinent to refer first briefly to the life cycle of *A. cantonensis* in rats since the complexity of the immune response is generated by the different developmental phases of the parasite. Moreover the adaptation and survival of *Angiostrongylus* in immune rats is a function of its biology.

Development of *A. cantonensis* in Rats

The life history of *A. cantonensis* in rats was elucidated by Mackerras and Sandars (1955); it is indirect and involves two hosts. Development to the third-stage larva takes place in an invertebrate, usually a mollusc. Rats, the definitive hosts, are infected by eating infected intermediate hosts. The third-stage larvae exsheath in the stomach, penetrate the tissues, and enter the blood stream. The larvae are then transported via the heart to the brain within 5 h after infection to congregate in the grey matter of the cerebral hemispheres, from where they gradually spread to other parts of the brain without causing much haemorrhage or reaction (Bhaibulaya 1975). Larvae moult to the fourth stage after 7 days and to the fifth stage betwen 9 and 11 days after infection. The young adult worms then move to the subarachnoid space and grow. Between 28 and 36 days after infection the worms migrate to the lungs via the venous system and mate soon after they reach the pulmonary arteries. Oviposition takes place between 40 and 50 days after infection. Parasite eggs are lodged in the capillaries of alveoli. Emboli form around them but the eggs hatch after 6 days. First-stage larvae migrate up the trachea to be swallowed and passed out with the faeces. They are immediately infective to the intermediate host.

Angiostrongyliasis is a chronic disease in rats. Adult worms survive and produce eggs for at least one year (Yong & Dobson 1982a).

Stimulus for Resistance

Heyneman and Lim (1965) first reported that rats actively acquire protective immunity to *A. cantonensis*. Numerous studies have since been undertaken to establish how many third-stage larvae are necessary to induce protective resistance in rats. These investigations were stimulated by the curious epidemiological observation that only small numbers of *A. cantonensis* are harboured by wild rats despite a wide range of infected intermediate host species and the high infectivity of this parasite for rats (Weinstein, Rosen, Laqueur & Sawyer 1963; Lim, Ow-Yang & Lie 1965; Yong, Welch & Dobson 1981). It was found that infections with more than 100 third-stage larvae are generally lethal to previously uninfected rats (Lim et al. 1965; Sirisinha, Techasoponmani, Dharmkrong-at & Uahkowithchai 1977; Yong & Dobson 1982a). This has led to the belief that wild rats that survive infection with *A. cantonensis* ingested small numbers of larvae in a manner analagous to experimental trickle infections. Moreover such small doses of larvae (5–50 third-stage larvae) given in a single administration can sensitize and protect rats against doses of larvae that are lethal in naive hosts (Heyneman & Lim 1965; Au & Ko 1979; Yong & Dobson 1982b).

The timing of the challenge infection appears to influence the intensity of reactions in sensitized rats. Au and Ko (1979) found that rats were protected against a challenge infection as early as 7 days after an immunizing infection but the immune reactions were more intense when the sensitized animals were challenged after 30 days. Lim and Heyneman (1969) also found that 2 doses of 10 third-stage larvae given at an interval of 2 weeks was more protective against a lethal challenge infection than a single sensitizing dose of 50 larvae. However, 2 sensitizing doses of 50 third-stage larvae did not significantly enhance the level of protective immunity acquired by rats given a single dose of 50 larvae. Instead, the quality rather than the degree of protection was changed and fewer male worms were recovered from the rats immunized twice (Yong & Dobson 1982c). In all cases the degree of protective immunity acquired by rats was not absolute.

All developmental stages of *A. cantonensis* stimulate resistance in rats but the intracranial larvae are more immunogenic than the intrapulmonary stages (Yoshimura, Aiba & Oya 1979). Lee (1969) effectively demonstrated this high larval immunogenicity when he rendered rats refractory to challenge infections by exposure to irradiated *A. cantonensis* third-stage larvae which were incapable of further development.

Immunity and the Host-Parasite Relationship

The dynamics of antibody response in rats after *A. cantonensis* infection reflect the development of the parasite and the induction of protective immunity in the host (table 8.1).

Table 8.1 Various immune responses detected in rats after *Angiostrongylus cantonensis* infection and their possible roles in the host-parasite relationship

Immune response	Role
1. Humoral immunity	
(a) Haemagglutinating antibody	Host protection
(b) Precipitating antibody	Parasite evasion (immune complexes)
(c) Complement fixing antibody	N.D.*
(d) Fluorescent conjugated antibody	Uncertain
(e) Percutaneous anaphylaxis (IgE)	Parasite evasion (suppressed and transient production)
2. Cell-mediated immunity +	
(a) *In vitro* blastogenic response	Host protection
(b) Macrophage migration inhibition	Host protection
(c) Delayed type skin reaction	Host protection

* N.D. not done. There is no record in the literature that the complement fixation test has ever been used in this infection.

\+ No details are available regarding the T cell subsets responsible for the reactions detected by each technique. In general, passive cell transfer experiments have shown that CMI confers protection.

Precipitating antibodies have been detected against crude adult worm antigen as early as 1 week after infection, reaginic antibodies after 2 weeks, and haemagglutinating antibodies after 4 weeks (Kamiya & Tanaka 1969; Kamiya, Tharavanij & Harinasuta 1973; Chen 1974; Chen & Susuki 1974; Yoshimura & Soulsby 1976; Dharmkrong-at, Uahkowithchai & Sirisinha 1978; Au & Ko 1979; Yoshimura, Aiba & Oya 1979; Yong & Dobson 1982d). In general, correlations between the levels of haemagglutinating antibodies and the development of the parasites and between these antibodies and the degree of resistance in infected rats are better than similar analyses using titres of antibodies assayed by different methods.

Dharmkrong-at et al. (1978) found anti-larval *A. cantonensis* haemagglutinating antibodies in serum of infected rats as early as 2 weeks after infection. These reached a peak titre after 4 weeks and disappeared by the 6th week after infection. The disappearance of these antibodies coincided approximately with the appearance of haemagglutinating antibodies to adult worm antigen. Rats which had either only adult female *A. cantonensis* or a mixture of both adult sexes implanted intraperitoneally produced haemagglutinating antibodies in their sera by 4 days after worm transfer, whereas antibodies were not detected in the sera of other rats which received only male worms until 5 weeks after transfer (Kamiya & Klongkamnuankarn 1970; Kamiya, Klongkamnuankarn, Tharavanij & Tanaka 1972). They concluded that the serum haemgglutinating antibody titres against *A. cantonensis* related to the maturation of adult female worms and to oviposition.

Reaginic antibodies are produced only transiently in rats after *A. cantonensis* infection and there is little or no anamnestic response to reinfection (Yoshimura & Yamagishi 1976). Yong and Dobson (1982a) also found that, unlike hamagglutinating antibody titres, the reaginic antibody response was related neither to the dose of larvae nor to the duration and number of infections. Allergic reactions accompanied by localized inflammation therefore do not appear to be as important in protective immune reactions in rats against *A. cantonensis* as has been found for other metazoa, particularly gastrointestinal parasites (Stewart 1955).

Cell-mediated immune (CMI) responses have been implicated in the immunity of rats to *A. cantonensis* from studies on the uptake of tritiated thymidine by antigen stimulated sensitized lymphocytes *in vitro* (Yoshimura & Soulsby 1976), and by the production of macrophage migration inhibition factor (Yoshimura, Aiba, Hayasaki & Yoshida 1976; Dharmkrong-at et al. 1978). Delayed-type hypersensitivity reactions have also been demonstrated in rats against *A. cantonensis in vivo* and the results suggest that these reactions parallel those obtained from *in vitro* tests (Dharmkrong-at et al. 1978). Elevated lung lysophospholipase activity and bone marrow eosinophilia (which are T cell regulated processes) provide additional evidence that CMI responses are induced during *A. cantonensis* infection in rats (Ottolenghi, Weatherly, Kocan & Larsh 1977; Laubach, Kocan & Sartain 1978). Furthermore, immunopathological changes suggestive of CMI responses such as granulomata around dead worms, and a heavy infiltration of lymphoid cells, neutrophil leucocytes and large mononuclear cells in the gastric mucosa, arteries, bronchi and in the adventitia of pulmonary veins are frequently observed in infected animals (see Alicata 1965).

The complexity of the *A. cantonensis* life cycle gives rise to speculation concerning the stage-specificity of the immunity detected in the host. Invariably the CMI responses detected are localized in lymphoid centres close to the areas of parasite invasion and migration (Yoshimura & Soulsby 1976; Dharmkrong-at et al. 1978; Au & Ko 1979). These have been interpreted as specific responses to the early developmental phases of *A. cantonensis*. Yong and Dobson (1982c) recently confirmed this relationship with experiments on the adoptive transfer of protective immunity in rats with cells from different lymphoid tissues.

Various hypotheses have been advanced to explain how hosts recognize the relative foreignness of their parasites. Sprent (1962) proposed the term "adaptation tolerance", Damian (1962, *Journal of Parasitology* **48**:[2, Sect. 2] 16) used the term "antigen mimicry" and Dineen (1963a, 1963b) wrote of "fitness antigens" for parasite antigens which corresponded to host components, to analyse the evolutionary adaptations of parasites to the immune reactions of the host. Dineen (1963a, 1963b) also envisaged host-parasite associations in which a degree of immunological interaction occurred, but where the stimulus initiating these reactions was kept below a threshold level. Population thresholds have been demonstrated for primary and for multiple *A. cantonensis* infections in rats which indicate possible adaptations of the host and parasite to each other (Yong & Dobson 1983).

A. cantonensis surviving from multiple infections respond differently to the immune responses of the rat compared with worms from primary infections. Larvae developing in immune rats grew into bigger adult worms than adult worms of the same age from normal rats (fig. 8.1). Moreover, the growth of *A. cantonensis* from primary infections was enhanced by the immune responses to larvae from reinfections but this effect was not permanent because host immunity soon reasserted its control on the parasite populations (Yong & Dobson 1983). Each of these phenomena reflects possible evolutionary interactions between host and parasite.

Mechanisms of Acquired Resistance

The roles of humoral and cell-mediated immunity in acquired resistance in rats against *A. cantonensis* have been demonstrated by passive transfer experiments. Both immune lymphocytes and antiserum protected recipient rats against establishment of *A. cantonensis* but antiserum conferred greater protection than passively transferred immune lymphocytes (Yong & Dobson 1982e). Other passive transfer experiments using immune spleen cells corroborated these results (Yong, Glanville & Dobson 1983). However, neither passively transferred rat antibodies nor immune cells achieved the levels of protection acquired following active infection. Little is known about the antibody isotypes, other serum factors (particularly complement) or specific cell populations that are required to produce protection against *A. cantonensis* in the rat. The importance of at least two different IgG subclasses acting synergistically in protective immunity has been demonstrated by Mitchell, Rajasekariah and Rickard (1980) with *Taenia taeniaeformis* in mice and by Yong, Das and Dachlan (1983) with *Schistosoma mansoni* in hamsters. There is evidence that the protective immunoglobulin against *A. cantonensis* in rats is associated with 7S but

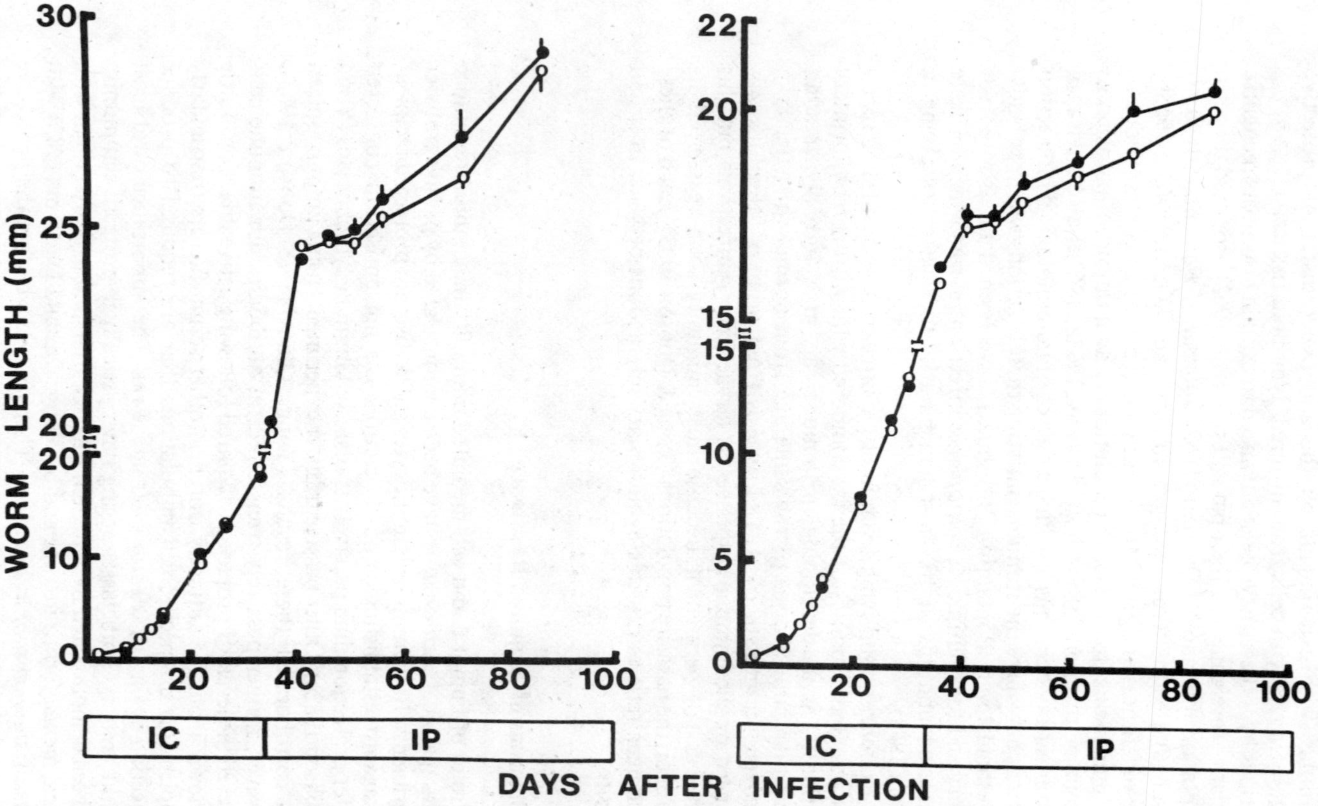

Figure 8.1 The growth of female (A) and male (B) *Angiostrongylus cantonensis* from primary (0) and secondary (●) infection in rats given 50 larvae. There was no difference in the worm length of larvae from primary and secondary infections during the intracranial (IC) stage, but significant differences were observed during the intrapulmonary (IP) stage of development in the life cycle of the parasite. Horizontal bar

not 19S fractions from immune rat serum (Kamiya, Klongkamnuankarn & Tharavanij 1972).

The protective antigens that induce humoral and CMI responses in rats have not yet been identified but Bouthemy, Capron, Afchain and Wattre (1972) separated 25 antigenic components in adult *A. cantonensis* homogenates and 15 components in fourth-stage larval extracts by immunoelectrophoresis. Common or cross-reactive antigens from different developmental stages have been found and these have included both somatic and metabolic components (Chen & Suzuki 1974; Techasoponmani & Sirisinha 1980). Metabolic antigens obtained from adult *A. cantonensis* maintained *in vitro* have been used to analyse the immune responses of infected animals (Jacobs, Lunde & Weinstein 1965, *Journal of Parasitology* **51**:[2, Sect. 2]38). Dharmkrong-at et al. (1978) found that antibody patterns obtained with metabolic antigens from adult female worms were similar to those of the somatic antigens from either male or female worms. However, no antibody activity was detected against metabolic antigens from adult male worms. Techasoponmani and Sirisinha (1980) have also found that metabolic products from adult female worms immunize rats and mice against *A. cantonensis* infection and suggested that female worms were the source of at least one of the protective antigens from the parasite.

Evasion of Immune Surveillance

A. cantonensis is long lived in the rat despite the development of effective host immunological surveillance mechanisms. In rats, the parasite is sensitive to immunological attack during the invasive stages of the larvae, but adult worms tolerate the immune response they engender (Heyneman & Lim 1965; Lim et al. 1965; Yong & Dobson 1983). This immunological interaction compares with the phenomenon of concomitant immunity described by Smithers, Terry and Hockley (1969) in schistosomiasis and later by others for different metazoan parasites (see Cohen 1976; Ogilvie & Worms 1976; Williams 1979). A number of attempts have been made to identify the mechanisms used by parasites to escape the effects of host protective immunity. Broadly speaking, the factors involved may be considered in relation to the host and alternately to the parasite (table 8.2).

Properties of the host

The capacity of the host to recognize antigens is a property of the lymphocyte-macrophage group of cells and it is determined genetically. These genetic regulatory mechanisms also influence host-parasite relationships.

Table 8.2 Factors permitting the escape of *Angiostrongylus cantonensis* from immunological destruction

1. Properties of the host
   (a) Absence of an effective immune response (genetic control)
   (b) Inappropriate immune response associated with the production of antibody, immune complexes, and regulatory cells
2. Properties of the parasite
   (a) Development in anatomically secluded site
   (b) Inertness of the worm cuticle
   (c) Weak immunogenicity
   (d) Secretion of immunosuppressive substances

There are, for example, numerous references to variations between strains and breeds of animals in their susceptibility and immune responses to helminth infections. The mechanisms underlying this variability are believed to be controlled by histocompatibility-linked immune response (Ir) genes, although other factors not associated with the major histocompatibility gene complex may also play an important role (see Ogilvie & Wilson 1976; Bloom 1979). Such genetically controlled factors are known to influence both the acquired resistance and the immune responses of rats to *A. cantonensis* infection (Yoshimura, Aiba, Hirayama & Yosida 1979). Strains of inbred rats that resist infection also produce significantly different patterns and higher levels of antibody than more susceptible strains. Little is known about the identity of the genes that control *A. cantonensis* infection in rats but their absence from rats correlates with enhanced survival of the parasite.

The function of the immune system of the host is complex. It is stimulated by antigens and augmented by elaborate interactions between cells and chemicals which govern its effector reactions. Parasites may prejudice the efficient activity of the immune response. For example, immunoglobulin production may go awry during parasite infections. In the case of *A. cantonensis* high concentrations of parasite specific immunoglobulins can be detected in infected rats which, more often than not, fail to relate to protective immunity. On the contrary, these immunoglobulins may contribute to parasite survival by jamming the effector arms of the immune response (Bloom 1979).

The continued survival of *A. cantonensis* in immune rats may relate to the sequestration of protective antibodies on the surface of specifically sensitized cells. Yong and Dobson (1982e) found that passive transfer of sensitized lymphocytes together with immune serum reduced the level of protection afforded to recipient rats by immune serum alone against a challenge with *A. cantonensis* larvae.

Rats infected with *A. cantonensis* fail to produce significant levels of reaginic antibody (Yoshimura & Soulsby 1976; Yoshimura & Yamagishi 1976; Yong & Dobson 1982d). This differs from many other metazoan parasite-host systems, where elevated IgE levels are signal diagnostic features, and allergic reactions accompanied by localized inflammation and eosinophilia are important protective reactions. Thus suppression of reaginic antibody production in infected rats may be advantageous to the survival of *A. cantonensis*.

IgE production in mammals is thymus dependent and is normally regulated by suppressor T cells (see Ishizaka 1976). It is not known whether *A. cantonensis* infection can activate such regulatory cells, but other parasites have been shown to have this capacity (Ljungstrom & Huldt 1977). Nevertheless, parasite induced depression of T cell reactivity (monitored by responsiveness to phytohaemagglutinin) and the presence of specific immune complexes which can activate suppressor T cells have been detected in rats during *A. cantonensis* infections (Yamashita, Saito, Sato, Takai, Watanabe & Otsuru 1979; Takai, Sato, Watanabe, Otsuru & Yamashita 1979; Gershon, Mokyr & Mitchell 1974). The activation of regulatory cells and the production of immunoregulatory complexes may therefore assist *A. cantonensis* to escape immunological destruction.

Properties of the parasite

Angiostrongylus cantonensis differs from many other parasites because it migrates to the brain of its host. The brain is an immunological haven (Burnet 1962), thus there would appear to be advantages for parasite survival in this environment. It is, however, teleological to argue that *A. cantonensis* has evolved a particular form of migratory behaviour to avoid the immune responses of the host. More probably the larvae reach the brain because they are driven there on their journey through the heart. Despite this the brain undoubtedly has been important in the evolution of the life cycle and as a consequence to the survival of *A. cantonensis*.

Kocan (1974) found that protective immunity was not evident when infected rats were challenged by surgical transplantation of larvae into the meninges, compared with animals challenged by mouth with infective larvae. Intrameningeal transplantion bypassed the normal migratory pathway followed by *A. cantonensis* larvae where many are destroyed in immune rats before they can reach the refuge of the brain. Moreover those larvae which survived migration from the stomach to the brain in immune rats after oral infection developed to young adults capable of surviving the debilitating effects of immunity in the lungs of the rat. These adapted worms were, nevertheless, structurally identical to those which develop following primary infections in rats. Thus the adaptation may, in part, have a physiological basis which is demonstrated by the enhanced growth of these survivors compared with worms from primary infections (fig. 8.1).

Several other lines of evidence are available which help to explain how adult *A. cantonensis* escape immunological destruction. Adult *A. cantonensis* have an immunological inert cuticle which may serve as a barrier to immunological attack (Chen & Suzuki 1974). Adult *A.*

cantonensis are less immunogenic than their larval stages. Parasite immunogenicity is reduced during larval development in the brain (Yoshimura, Aiba & Oya 1979). This change is demonstrated by the experiments of Ko (1979). He found that young fifth-stage *A. cantonensis* recovered from the subarachnoid space of rats develop normally after intracranial transplantation into recipient rats, but similar worms from mice do not. However, when third- and fourth-stage larvae from mice are transplanted into the heads of recipient rats they develop to normal adult parasites (Ko 1979). Thus young *A. cantonensis* change their antigens at, or soon after, their final moult. Indeed, "adapted" young adult worms from immune rats provoke little or no cellular reactions when transplanted into naive recipient rats. In contrast, recipient animals other than rats not only killed similarly transplanted "adapted" worms but they were provoked to mount massive cellular reactions and specific antibody responses against these worms (Ko 1979; Yoshimura, Aiba, Oya & Fukuda 1980). Thus "adapted" young adult worms appear less immunogenic in rats but they are not so in other host species. It is not known whether the adapted worms acquire host components or whether they change their own composition to avoid recognition. Several recent reviews and a symposium have been devoted to explaining how other mature worm species survive damage by immunological reactions (Porter & Knight 1974; Ogilvie & Wilson 1976; Bloom 1979).

Helminth parasites appear to react antagonistically to immunological attack. Their antagonism may relate to the establishment of microenvironments that prejudice the efficient functions of the effector cells and chemicals involved in resistance. For example, *A. cantonensis* may subvert immunological attack by shedding immunosuppressive secretions and antigens that block protective immune reactions.

Free soluble antigens (Chen, Suzuki & Liu 1973) and immune complexes circulate in rats infected with *A. cantonensis* (Yamashita et al. 1979; Takai et al. 1979). Similar substances are believed to depress host immunity against other parasitic infection through antigen competition and immune deviation (Wilson 1974). These reactions result in the loss of high affinity protective antibodies bound in complexes with circulating antigens so that only antibodies with low affinity and low protective efficacy against the parasites remain.

Helminth parasites secrete enzymes, some of which induce protective immunity (see Soulsby 1963; Clegg & Smith 1978), whereas others may act to hinder these responses. *A. cantonensis* synthesizes large quantitites of acetylcholinesterase which varies in its isoenzyme pattern as the parasite develops (Beaver & Dobson 1978). Changes in the quantity and quality of acetylcholinesterase in normal and immunologically-damaged and -adapted *Nippostrongylus brasiliensis* also occur (Edwards, Burt & Ogilvie 1971). The significance of these enzyme changes is not clear but the production of acetylcholinesterase in large quantity implies that it has an important function in the biology of these parasites. Beaver and Dobson

(1978) suggested that the secretion of acetylcholinesterase isoenzyme variants assisted *A. cantonensis* to avoid immunological damage by compromising the immune responses of the host.

Conclusion

The host-parasite interaction embodies a struggle for survival between two organisms in which the host antagonizes parasite establishment and to which, in turn, the parasite responds and adapts.

Parasites appear to have evolved mechanisms and responses to either interfere with or avoid the immune reactions they induce in their hosts. These adaptations may arise at a particular time in the life cycle of the parasite. For *A. cantonensis* these changes occur in the brain of the rat as the larva moults and becomes a young adult. It is intriguing that *A. cantonensis* makes these adaptations while it is in an immunological haven just before it re-enters the immunologically hostile circulation.

The evolutionary process governing such host-parasite interactions is summarized in the "gene-for-gene" concept which states that host and parasite develop complementary genic systems, and in the notion of coevolution (see Day 1974; May & Anderson 1982). The longevity of *A. cantonensis* infections in rats illustrates the potential of one nematode species to survive despite the adverse effects of host immunity.

Acknowledgments

We thank Drs G.G. Riffkin, C. Lenghaus, J.J. Webber and A.P.L. Callinan for reading and criticizing the manuscript, Miss E.K. Aldridge for preparing the figure, and Miss W.C. Huf and Mrs S. Roth for their excellent typing.

References

Alicata, J.E. 1965. Biology and distribution of the rat lungworm *Angiostrongylus cantonensis*, and its relationship to eosinophilic meningoencephalitis and other neurological disorders of man and animals. *Advances in Parasitology* **3**:223–48.

Au, A.C.S. and Ko, R.C. 1979. Changes in worm burden, haematological and serological response in rats after single and multiple *Angiostrongylus cantonensis* infections. *Zeitschrift für Parasitenkunde* **58**:233–42.

Beaver, J.A. and Dobson, C. 1978. Acetylcholinesterase levels in *Angiostrongylus cantonensis* in relation to the immune response in rats. *International Journal for Parasitology* **8**:9–13.

Bhaibulaya, M. 1975. Comparative studies on the life history of *Angiostrongylus mackerrasae* Bhaibulaya, 1968 and *Angiostrongylus cantonensis* (Chen, 1935). *International Journal for Parasitology* **5**:7–20.

Bloom, B.R. 1979. Games parasites play: How parasites evade immune surveillance. *Nature (London)* **279**:21–26.

Bouthemy, F., Capron, A., Afchain, D. and Wattre, P. 1972. Antigenic structure of the nematode *Angiostrongylus cantonensis*. Immunologic aspects of host-parasite relationship. *Annales de Parasitologie Humaine et Comparée* **47**:531–50.

Burnet, F.M. 1962. *The integrity of the body*. London: Oxford University Press.

Chen, S.N. 1974. Studies on immunodiagnosis of angiostrongyliasis. 5. Double immunodiffusion and counterelectrophoresis for detection of antibody in infected rats and human sera. *Chinese Journal of Microbiology* **7**:114–18.

Chen, S.N. and Suzuki, T. 1974. Fluorescent antibody and indirect haemagglutination tests for *Angiostrongylus cantonensis* infection in rats and rabbits. *Journal of the Formosan Medical Association* **73**:393–400.

Chen, S.N., Suzuki, T. and Liu, K.H. 1973. Studies on immunodiagnosis of angiostrongyliasis. 1. Detection of antigen and antibody in serum and cerebrospinal fluid. *Journal of the Formosan Medical Association* **72**:161–66.

Clegg, J.A. and Smith, M.A. 1978. Prospects for the development of dead vaccines against helminths. *Advances in Parasitology* **16**:165–218.

Cohen, S. 1976. Survival of parasites in the immunized host. In *Immunology of parasitic infections*, ed. S. Cohen and E.H. Sadun, 35–46. Oxford: Blackwell Scientific Publications.

Day, P.R. 1974. *Genetics of host-parasite interaction*. San Francisco: W.H. Freeman and Company.

Dharmkrong-at, A., Uahkowithchai, V. and Sirisinha, S. 1978. The humoral and cell-mediated immune responses to somatic and metabolic antigens in rats infected with *Angiostrongylus cantonensis*. *Southeast Asian Journal of Tropical Medicine and Public Health* **9**:330–37.

Dineen, J.K. 1963a. Immunological aspects of parasitism. *Nature (London)* **197**:268–69.

Dineen, J.K. 1963b. Antigenic relationship between host and parasite. *Nature (London)* **197**:471–72.

Dineen, J.K. and Wagland, B.M. 1982. Immunoregulation of parasites in natural host-parasite systems — with special reference to the gastrointestinal nematodes of sheep. In *Biology and control of endoparasites*, ed. L.E.A. Symons, A.D. Donald and J.K. Dineen, 297–329. New York: Academic Press.

Dobson, C. 1972. Immune response to gastrointestinal helminths. In *Immunity to animal parasites*, ed. E.J.L. Soulsby, 191–222. New York: Academic Press.

Edwards, A.J., Burt, J.S. and Ogilvie, B.M. 1971. The effect of immunity on some enzymes of the parasitic nematode *Nippostrongylus brasiliensis*. *Parasitology* **62**:339–47.

Gemmell, M.A. and MacNamara, F.N. 1972. Immune response to tissue parasites. II. Cestodes. In *Immunity to animal parasites*, ed. E.J.L. Soulsby, 236–72. New York: Academic Press.

Gershon, R.K., Mokyr, M.B. and Mitchell, M.S. 1974. Activation of suppressor T cells by tumour cells and specific antibody. *Nature (London)* **250**:594–96.

Heyneman, D. and Lim, B.L. 1965. Prolonged survival in rats immunized by a small number of low-level doses of *Angiostrongylus cantonensis* and challenged with a lethal level of infective larvae. *Medical Journal of Malaya* **20**:162–63.

Ishizaka, K. 1976. Cellular events in the IgE antibody response. *Advances in Immunology* **23**:1–75.

Kamiya, M. and Klongkamnuankarn, K. 1970. Haemagglutination (HA) activity after the transfer of adult worms of *Angiostrongylus cantonensis* to the abdominal cavity of normal rats. *Southeast Asian Journal of Tropical Medicine and Public Health* **1**:571–72.

Kamiya, M., Klongkamnuankarn, K. and Tharavanij, S. 1972. Fractionation study of haemagglutination antibodies in rats infectd with *Angiostrongylus cantonensis. Southeast Asian Journal of Tropical Medicine and Public Health* **3**:397–402.

Kamiya, M., Klongkamnuarnkarn, K., Tharavanij, S. and Tanaka, H. 1972. Change of indirect haemagglutination reactions in serum after the transfer of adult *Angiostrongylus cantonensis* to the abdominal cavity of rats. *Southeast Asian Journal of Tropical Medicine and Public Health* **3**:119–23.

Kamiya, M. and Tanaka, H. 1969. Haemagglutination test in rats infected with *Angiostrongylus cantonensis. Japanese Journal of Experimental Medicine* **39**:593–99.

Kamiya, M., Tharavanij, S. and Harinasuta, C. 1973. Antigenicity for haemagglutination and immunoelectrophoresis tests in fractionated antigens from *Angiostrongylus cantonensis. Southeast Asian Journal of Tropical Medicine and Public Health* **4**:187–94.

Ko, R.C. 1979. Host-parasite relationship of *Angiostrongylus cantonensis*. 1. Intracranial transplantation into various hosts. *Journal of Helminthology* **53**:121–26.

Kocan, A.A. 1974. The influence of *Nippostrongylus brasiliensis* on the establishment of *Angiostrongylus cantonensis* on the laboratory rat. *Proceedings of the Helminthological Society of Washington* **41**:237–41.

Laubach, H., Kocan, A.A. and Sartain, K.E. 1978. Effects of various numbers of adult *Angiostrongylus cantonensis* on lung lyophospholipase activities and bone marrow eosinophil levels of specific pathogen-free rats. *Journal of Parasitology* **64**:1145–46.

Lee, S.H. 1969. The use of irradiated third-stage larvae of *Angiostrongylus cantonensis* as antigen to immunize albino rats against homologous infection. *Proceedings of the Helminthological Society of Washington* **36**:95–97.

Lim, B.L. and Heyneman, D. 1969. Further cross-infection studies of three strains of *Angiostrongylus cantonensis* (Nematode:Metastrongylidae). *Proceedings of seminar on filariasis and immunology of parasitic infections and laboratory meeting* Singapore 1968, pp.68–72.

Lim, B.L., Ow-Yang, C.K. and Lie, K.J. 1965. Natural infection of *Angiostrongylus cantonensis* in Malayan rodents and intermediate hosts, and preliminary observations of acquired resistance. *American Journal of Tropical Medicine and Hygiene* **14**:610–17.

Ljungstrom, I. and Huldt, G. 1977. Effect of experimental trichinosis on unrelated humoral and cell-mediated immunity. *Acta Pathologica et Microbiologica Scandinavica Section C: Immunology* **85**:131–41.

Mackerras, M.J. and Sandars, D.F. 1955. The life-history of the rat lungworm, *Angiostrongylus cantonensis* (Chen) (Nematode:Metastrongylidae). *Australian Journal of Zoology* **3**:1–21.

May, R.M. and Anderson, R.M. 1982. Coevolution of hosts and parasites. *Parasitology* **85**:411–26.

Mitchell, G.F., Rajasekariah, G.R. and Rickard, M.D. 1980. A mechanism to account for mouse strain variation in resistance to the larval cestoda, *Taenia taeniaeformis. Immunology* **39**:481–89.

Ogilvie, B.M. and Rose, M.E. 1978. The response of the host to some parasites of the small intestine: Coccidia and nematodes. *Les Colloques de L'Institut National de la Santé et de la Recherche Médicale — Immunity in Parasitic Diseases*. INSERM, September 1977, Vol. 72, pp.237–248.

Ogilvie, B.M. and Wilson, R.J.M. 1976. Evasion of the immune response by parasites. *British Medical Bulletin* **32**:177–81.

Ogilvie, B.M. and Worms, M.J. 1976. Immunity to nematode parasites. In *Immunology of parasitic infections*, ed. S. Cohen and E.H. Sadun, 380–407. Oxford: Blackwell Scientific Publications.

Olson, L.J. and Izzat, N.N. 1972. Immune response to tissue helminths. I. Nematodes. In *Immunity to animal parasites*, ed. E.J.L. Soulsby, 223–34. New York: Academic Press.

Ottolenghi, A., Weatherly, N.F., Kocan, A.A. and Larsh, J.E. Jr. 1977. *Angiostrongylus cantonensis*: Phospholipase in nonsensitized and sensitized rats after challenge. *Infection and Immunity* **15**:13–18.

Porter, R. and Knight, J. 1974. Eds. *Ciba Foundation Symposium 25 (Parasites in the immunized host: Mechanisms of survival)*. Amsterdam: Associated Scientific Publishers.

Sirisinha, S., Techasoponmani, R., Dharmkrong-at, A. and Uahkowithchai, V. 1977. Immunology of *Angiostrongylus* infection. *Journal of the Science Society of Thailand* **3**:157–74.

Smithers, S.R. and Terry, R.J. 1976. The immunology of schistosomiasis. *Advances in Parasitology* **14**:399–422.

Smithers, S.R., Terry, R.J. and Hockley, D.J. 1969. Host antigens in schistosomiasis. *Proceedings of the Royal Society. Series B* **171**:483–94.

Soulsby, E.J.L. 1963. The nature and origin of the functional antigens in helminth infections. *Annals of the New York Academy of Sciences* **113**:492–506.

Soulsby, E.J.L., Monsell, G. and Lloyd, S. 1981. Acquisition of immunological competence to gastro-intestinal trichostrongyles by young ruminants: Epidemiological significance. *Current Topics in Veterinary Medicine and Animal Science* **9**:513–26.

Sprent, J.F.A. 1962. Parasitism, immunity and evolution. In *The evolution of living organisms*, ed. G.W. Leeper, 149–65. Parkville, Victoria: Melbourne University Press.

Stewart, D.F. 1955. "Self-cure" in nematode infections of sheep. *Nature (London)* **176**:1273–74.

Takai, A., Sato, Y., Watanabe, H., Otsuru, M. and Yamashita, T. 1979. Studies on the circulating immune complexes in rats infected with *Angiostrongylus cantonensis*. Detection of antigen in the immune complexes precipitated by PEG. *Japanese Journal of Parasitology* **28**:411–20.

Techasoponmani, R. and Sirisinha, S. 1980. Use of excretory and secretory products from the adult female worms to immunize rats and mice against *Angiostrongylus cantonensis* infection. *Parasitology* **80**:457–69.

Weinstein, P.P., Rosen, L., Laqueur, G.L. and Sawyer, T.K. 1963. *Angiostrongylus cantonensis* infection in rats and rhesus monkeys, and observations on the survival of the parasite *in vitro*. *American Journal of Tropical Medicine and Hygiene* **12**:358–77.

Williams, J.F. 1979. Recent advances in the immunology of cestode infections. *Journal of Parasitology* **65**:337–49.

Wilson, R.J.M. 1974. Soluble antigens as blocking antigens. In *Ciba Foundation Symposium 25 (Parasites in the immunized host: Mechanisms of survival)*, ed.

R. Porter and J. Knight, 185–95. Amsterdam: Associated Scientific Publishers.

Yamashita, T., Saito, Y., Sato, Y., Takai, A., Watanabe, H. and Otsuru, M. 1979. Circulating antigens and immune complexes in the serum of rat infected with *Angiostrongylus cantonensis. Japanese Journal of Parasitology* **28**:393–401.

Yong, W.K. and Dobson, C. 1982a. Population dynamics of *Angiostrongylus cantonensis* during primary infections in rats. *Parasitology* **85**:399–409.

Yong, W.K. and Dobson, C. 1982b. The biology of *Angiostrongylus cantonensis* in immune rats. *Southeast Asian Journal of Tropical Medicine and Public Health* **13**:244–48.

Yong, W.K. and Dobson, C. 1982c. The passive transfer of protective immunity against *Angiostrongylus cantonensis* with lymph node cells from different lymphoid tissues in rats. *International Journal for Parasitology* **12**:423–25.

Yong, W.K. and Dobson, C. 1982d. Antibody response in rats infected with *Angiostrongylus cantonensis* and the passive transfer of protective immunity with immune serum. *Zeitschrift für Parasitenkunde* **67**:329–36.

Yong, W.K. and Dobson, C. 1982e. Passive immunity in rats infected with *Angiostrongylus cantonensis*: Interaction between syngeneic immune serum and sensitized lymph node cells. *Zeitschrift für Parasitenkunde* **68**:87–92.

Yong, W.K. and Dobson, C. 1983. Immunological regulation of *Angiostrongylus cantonensis* infections in rats: Modulation of population density and enhanced parasite growth following one or two superimposed infections. *Journal of Helminthology* **57**:155–65.

Yong, W.K., Welch, J.S. and Dobson, C. 1981. Localized distribution of *Angiostrongylus cantonensis* among wild rat populations in Brisbane, Australia. *Southeast Asian Journal of Tropical Medicine and Public Health* **12**:608–9.

Yong, W.K., Das, P.K. and Dachlan, Y.P. 1983. *Schistosoma mansoni* infection of Syrian golden hamsters: The host humoral immune response in relation to the adult worm burdens after primary infection. *Zeitschrift für Parasitenkunde* **69**:41–51.

Yong, W.K., Glanville, R.J. and Dobson, C. 1983. The role of the spleen in protective immunity against *Angiostrongylus cantonensis* in rats: Splenectomy and passive spleen cell transfers. *International Journal for Parasitology* **13**:165–170.

Yoshimura, K. and Soulsby, E.J.L. 1976. *Angiostrongylus cantonensis*: Lymphoid cell responsiveness and antibody production in rats. *American Journal of Tropical Medicine and Hygiene* **25**:99–107.

Yoshimura, K. and Yamagishi, T. 1976. Reagenic antibody production in rabbits and rats infected with *Angiostrongylus cantonensis. Japanese Journal of Veterinary Science* **38**:33–40.

Yoshimura, K., Aiba, H. and Oya, H. 1979. Transplantation of young adult *Angiostrongylus cantonensis* into the rat pulmonary vessels and its application to the assessment of acquired resistance. *International Journal for Parasitology* **9**:97–103.

Yoshimura, K., Aiba, H., Hirayama, N. and Yoshida, T.H. 1979. Acquired resistance and immune responses of eight strains of inbred rats to infection with *Angiostrongylus cantonensis. Japanese Journal of Veterinary Science* **41**:245–59.

Yoshimura, K., Aiba, H., Hayasaki, M. and Yoshida, H. 1976. Delayed hypersensitivity responses of guinea pig and rat to *Angiostrongylus cantonensis* infection. *Japanese Journal of Veterinary Science* **38**:579–93.

Yoshimura, K., Aiba, H., Oya, H. and Fukuda, T. 1980. *Angiostrongylus cantonensis*: Development following pulmonary arterial transfers into permissive and nonpermissive hosts. *Experimental Parasitology* **49**:339–52.

9 Mechanisms Involved in the *in vitro* Killing of *Dirofilaria immitis* Microfilariae*

C.M. Rzepczyk

Introduction

The factors involved in the clearance of circulating *Dirofilaria immitis* microfilariae have not yet been fully characterized. Even under controlled experimental conditions, it is not understood why some dogs develop a persistent microfilaraemia when infected with infective larvae, whereas others show only a transient microfilaraemia, and still others do not develop a patent microfilaraemia at all (Weil, Powers, Parbuoni, Line, Furrow & Ottesen 1982). Nevertheless, various studies have suggested that the immune status of the host is involved, as the amicrofilaraemic state has been associated with the presence of antibodies to microfilarial antigens (Rawlings, Dawe, McCall, Keith & Prestwood 1982) and can be produced experimentally by immunization (Wong, Suter, Rhode & Guest 1973).

Evidence that an antibody-dependent cell-mediated immune response may be involved in the development of the amicrofilaraemic state comes primarily from studies on other filarial species (Mehta, Sindhu, Subrahmanyam, Hopper & Nelson 1980; Mehta, Sindhu, Subrahmanyam, Hopper, Nelson & Rao 1981; Haque, Ouassi, Joseph, Capron & Capron 1981). These and other studies have shown that the destruction of microfilariae *in vitro* occurs only in the presence of an appropriate antibody with or without complement, and effector cells. The inference that such cell-mediated killing operates also *in vivo* is important as it may explain much of the pathology, e.g., the lymphatic and pulmonary lesions (Klei, Enright, Blanchard & Uhl 1982; Wong & Guest 1969; Castleman & Wong 1982) seen in chronic filarial infections characterized by amicrofilaraemia. This study was initiated to look at the factors required for cellular adhesion and cytotoxicity to *D. immitis* microfilariae *in vitro*.

Materials and Methods

RPMI 1640 (Gibco, Diagnostics, Madison, Wis.) culture medium supplemented per litre with 5.96 g HEPES (Sigma Chemical Co., St.

* This work was conducted in accordance with the NHMRC/CSIRO Code of Practice for the care and use of animals in research in Australia.

Louis, Mo.), 40 mg gentamyacin sulphate (Schering Corporation USA Kenilworth, NJ) and 10% heat-inactivated foetal calf serum was used in all instances except for the filtering and washing of microfilariae for which serum-free medium was used.

Dog sera were prepared by allowing blood to clot for 1 h at room temperature. The sera were separated, filtered and then stored in small aliquots at $-70°C$. Heat-inactivation of serum samples (56°C, 30 min) was done just prior to use. All sera were screened for occult dirofilariasis using an indirect fluorescent antibody test (IFA–mf Slide Test for Dirofilariasis, Immunovetcal Inc., San Francisco). Using fluorescein isothiocyanate-labelled rabbit anti-canine gamma globulin, this test detects specific antibodies against cuticular and somatic antigens of *D. immitis* microfilariae. In a positive test fluorescence is seen at both cuticular and somatic sites of the microfilariae.

Sera were obtained from normal, microfilaraemic and amicrofilaraemic (occult) dogs. The occult infections (dogs OC1–OC4) were diagnosed clinically and radiographically (Carlisle 1980) at the Veterinary Clinic of the University of Queensland. In dog OC3 this diagnosis was confirmed at post-mortem. Microfilaraemic dogs (MF1–MF5) were obtained from a local pound: their microfilarial concentrations ranged from 3.5×10^2 – 2×10^4/ml. The two normal dogs (ND1 and ND2) had been kept in a screened animal house for most of their lives. Microfilaraemias were never detected in these dogs and they were negative for occult dirofilariasis by the IFA–mf Slide Test (Immunovetcal Inc.) described above.

Two dogs (MF3 and MF5) were used as donors of *D. immitis* microfilariae. Blood from both dogs tested negative for *Dipetalonema reconditum* microfilariae using an acid phosphatase stain (Chalifoux & Hunt 1971). MF3 and MF5 had microfilaraemias in the order of 3×10^3 and 2×10^4 per ml respectively. Microfilariae were recovered from freshly collected heparinized dog blood by filtration through 5 μm polycarbonate filters (Subrahmanyam, Mehta, Nelson & Rao 1978). The recovery process was repeated once to produce a suspension of microfilariae essentially free of contaminating host cells. Viability was greater than 95%. Microfilariae were re-suspended to a concentration of 2×10^3/ml and stored overnight in a CO_2 incubator at 37°C. They were washed once just prior to use.

Effector cells were obtained from dog ND2 and were prepared by sedimenting 10 ml of heparinized blood with 2 ml 6% Macrodex (Macrodex-dextran 70, Pharmacia) as described by Mehta et al. (1981). The erythrocytes were lysed with a hypotonic lysing solution (0.83% NH_4Cl in 0.017 M Trisma Base [Sigma], pH 7.2) at room temperature. After two washes, the leucocytes were counted and resuspended to a concentration of 1×10^6/ml. Viability as assessed by trypan blue staining was greater than 95%. Differential cell counts were determined from Giemsa-stained cytocentrifuge preparations. The proportion of cell types recovered approximated the numbers seen in peripheral blood. The mean values were 26% mononuclear cells, 66% neutrophils and 8% eosinophils. These cells were used in all assays except one in which purified neutrophils were also used. Neutrophils were prepared by taking the washed, unlysed cell preparations obtained by Macrodex sedimentation, and loading them onto a discontinuous metrizamide gradient as described by Vadas, David, Butterworth, Pisani & Siongok (1979). The cells in the two layers immediately above the erythrocyte pellet were harvested. They were washed and resuspended to 1 × 10^6/ml. Neutrophils of 95% purity were recovered. Viability as assessed by trypan blue staining likewise exceeded 95%.

All serum-dependent adhesion and cytotoxicity assays were carried out in 96 well flat-bottom microtitre plates (Nunclon, Delta, Denmark). The final volume of 100 μl was made up of 25 μl volumes each of medium, of normal or test dog serum, of microfilariae, and of effector cells. The test sera were used at a final concentration of 12.5%. They were diluted with either medium (table 9.1, experiments 1, 2 & 5) or with ND1 serum (table 9.1, experiments 2 & 4). Unless otherwise stated all cultures were set up in triplicate.

Plates containing medium, serum and microfilariae were incubated at 37°C for 30 min in an atmosphere of 5% CO_2 before the addition of effector cells. Examination for adhesion and cytotoxicity were performed after further 1 h and 16 h incubations using an inverted microscope (Olympus, Tokyo) at magnification × 200. Cellular adhesion was assessed qualitatively. Cytotoxicity was determined by counting the number of mobile and immobile microfilariae in each well. Worms immobile during the period of scoring were considered dead. These data were converted to show the percentage of dead microfilariae and were expressed as a mean of replicate samples. Cytocentrifuge preparations were made of the cultures at both the 1 h and 16 h time points. Material was fixed for electron microscopy by addition of 2% glutaraldehyde in 0.1 M cacodylate buffer (pH 7.3; 4°C), the osmolality of which had previously been adjusted to 300 mOsM with sucrose. Fixation time was 45 min.

Results

Examination of the cultures after 1 h incubation showed that the sera of all the occult dogs promoted the adhesion of peripheral blood leucocytes to the microfilariae. In the presence of non-heated occult dog sera, there was a generally dense and uniform coating of cells over the entire microfilarial surface (fig.9.1), whereas heated occult dog sera caused a variable degree of adherence with occasionally some microfilariae entirely free of cells (fig. 9.2). Cytotoxicity to microfilariae was not evident at this stage, but in the majority of experiments some microfilariae already showed impaired motility. The sera from normal dogs and

Table 9.1 Percentage of *D. immitis* microfilariae dead after 16 h incubation with canine peripheral blood leucocytes and serum from either occult, microfilaraemic or normal dogs[1]

	Experiment* 1		Experiment* 2		Experiment* 3**		Experiment* 4		Experiment* 5	
Serum***	Non-heated	Heated	Non-heated	Heated	Non-heated	Heated	Non-heated	Heated	Non-Heated	Heated
OC1	—	—	—	—	44	62	69	75	29	25
OC2	91	78	65	67	—	45	—	—	34	39
OC3	83	93	—	—	75	84	—	—	6	24
OC4	—	—	—	—	46	39	—	—	22	18
ND1	0	0	2	<1	3	0	0	0	0	1
ND2	0	0	<1	0	1	0	0	0	0	1
MF1	0	0	<1	1	—	—	—	—	2	2
MF2	0	0	<1	<1	—	—	—	—	1	1
MF3	0	0	<1	2	—	—	—	—	<1	2
MF4	0	0	3	1	—	—	—	—	2	2

* Blood microfilariae from dog MF3 were used in experiments 1 to 4. Dog MF5 was the microfilarial donor in experiment 5.

** Counts based on duplicate samples only

*** *OC* serum from dogs with occult infection; *ND* serum from uninfected dogs; *MF* serum from dogs with circulating microfilariae

[1] Part of the data given in this table has since been published in table 1 of Rzepczyk, C.M. & Bishop, C.J. (1984).

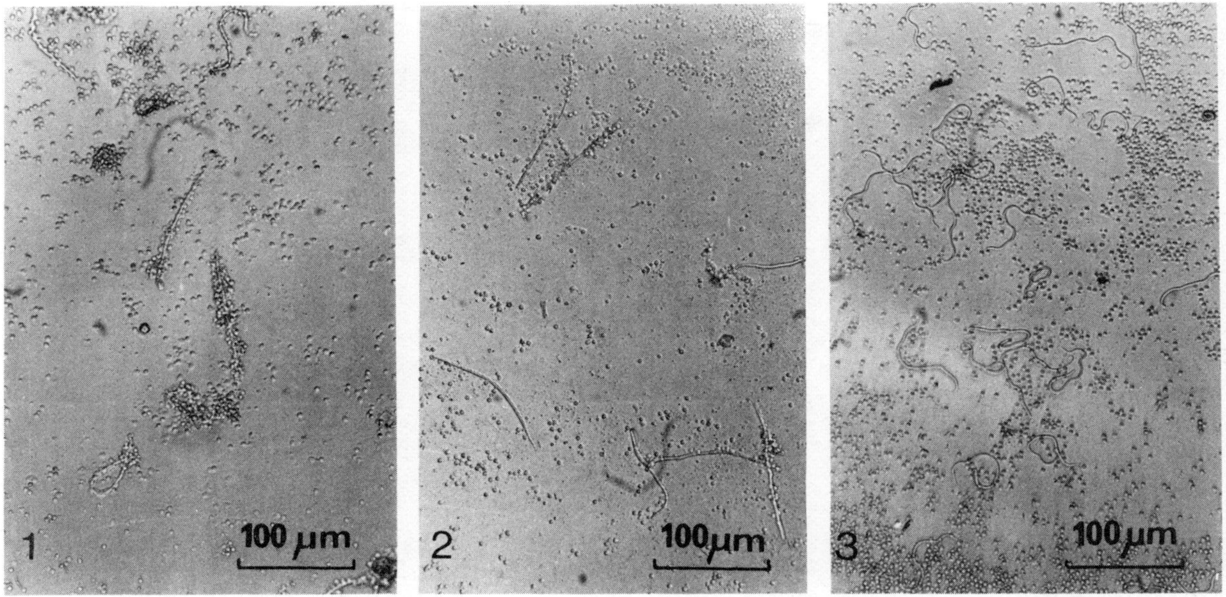

Figure 9.1 *D. immitis* microfilariae after 16 h incubation with mixed canine leucocytes and non-heated serum from an amicrofilaraemic (occult) dog. All the microfilariae are coated with cells and this cover is relatively uniform.

Figure 9.2 *D. immitis* microfilariae after 1 h incubation with mixed canine leucocytes and heat-inactivated serum from an amicrofilaraemic (occult) dog. The degree of cellular cover is variable, with some microfilariae free of cells.

Figure 9.3 *D. immitis* microfilariae after 1 h incubation with mixed canine leucocytes and non-heated serum from a normal dog. There is no adhesion of leucocytes to the worms. A similar result was obtained with the sera from microfilaraemic dogs.

microfilaraemic dogs did not promote adhesion (fig.9.3) and there was no evidence of impaired microfilarial motility.

After 16 h incubation, significant cytotoxicity was observed only in those cultures containing occult dog sera (table 9.1): killing occurred with heated as well as non-heated sera. The degree of cellular adhesion to microfilariae in these cultures reflected the findings at the earlier (1 h) time point, but there was now also some degeneration of effector cells, especially in cultures where there had been good killing of microfilariae. While the majority of microfilariae incubated for 16 h with occult dog sera were covered with cells and were either dead or had grossly impaired motility, there were some which were less severely affected. This apparent heterogeneity in the microfilarial population was particularly evident in some cultures containing heat-inactivated occult dog sera, where microfilariae, totally free of cells and highly active, were seen in association with dead or dying larvae well-coated with cells.

In some experiments, after 16 h there was also a limited degree of cellular adhesion to microfilariae in cultures containing normal or microfilaraemic dog sera. This adhesion was usually characterized by the attachment of a small rosette of cells to the anterior end of the larval worm. The adhesion was not associated with any impairment of the microfilariae and was largely abolished when heat-inactivated sera were used. Once again a degree of heterogeneity was observed in the susceptibility of the microfilariae to cellular attack, as occasionally, in cultures containing sera from normal or microfilaraemic dogs, microfilariae were seen with cells attached over their entire length. This cover of cells ranged from light to heavy. There was no evidence that just microfilariae already impaired or dead at commencement of incubation were subject to this adhesion, since dead and impaired microfilariae without any cellular cover were also present in these cultures. Furthermore, some very active microfilariae with a very heavy cover of cells were seen: the cells covering these worms did not appear degenerate when viewed under the inverted microscope. Such microfilariae did not exceed 5% of the total number.

The examination of cytocentrifuge preparations of cultures after 1 and 16 h of incubation indicated that neutrophils may be the primary effector cells involved in the adhesion and cytotoxicity to microfilariae. These cells were seen to be closely applied to the surface of the microfilariae even after 1 h incubation. This conclusion was supported by the results of an experiment with purified (= 95%) neutrophils which were capable of causing cytotoxicity to microfilariae in the presence of the serum from an occult dog (OC1). In this experiment, the mean percentage of dead microfilariae in parallel cultures in which neutrophils were the effector cells was 34% with non-heated serum, and 62% with heated serum which compared with 77% and 76% cytotoxicity respectively in cultures containing mixed leucocytes. There was no death or impairment of microfilariae in cultures containing dog sera but no effector cells.

The close association between neutrophils and the microfilarial surface was confirmed by electron microscopy (fig.9.4). The neutrophils appeared activated showing vacuolation and many cytoplasmic processes. These cytoplasmic processes were seen to extend into the indentations of the microfilarial cuticle (the cuticular crypts). Studies are in progress to determine if damage to the microfilariae is initiated at these sites.

The sera of all occult dogs and not the sera of any of the normal or microfilaraemic dogs gave a positive indirect flourescent antibody test. All the occult sera showed very strong fluorescence for both cuticular and somatic microfilarial surfaces.

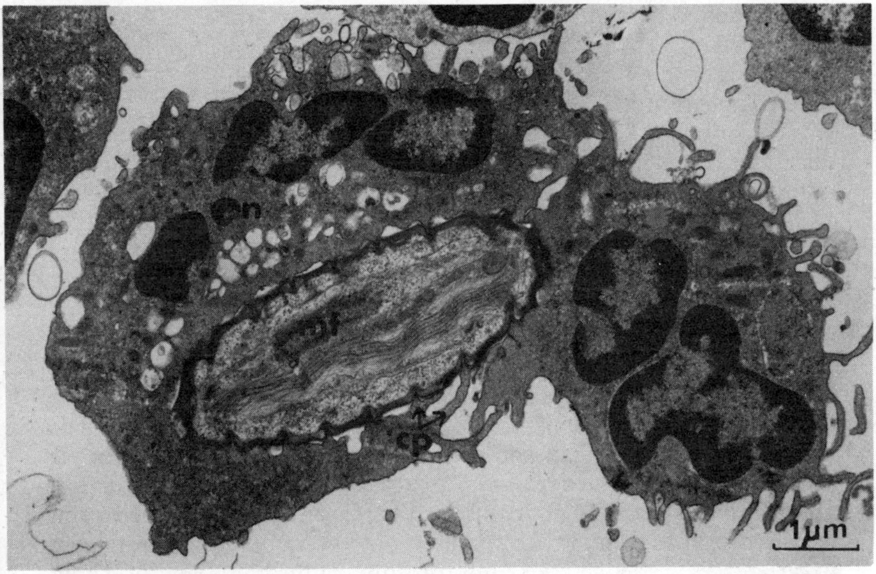

Figure 9.4 Electron micrograph showing the close association between the microfilariae and the effector cells (neutrophils) after 4 h incubation with the serum from an occult dog. Numerous cytoplasmic processes are seen to extend from the neutrophils into the indentations of the worm cuticle. *n* = neutrophil; *mf* = microfilarial worm; *cp* = cytoplasmic processes.

Discussion

This study demonstrated that the sera of amicrofilaraemic dogs, but not of microfilaraemic or uninfected dogs, promoted cellular adhesion and cytotoxicity to *D. immitis* microfilariae in an *in vitro* assay. This correlated with the detection in the sera of these amicrofilaraemic dogs of antibodies to cuticular and somatic microfilarial antigens. Heat-inactivation of these sera was shown to reduce the degree of cellular adherence to the microfilariae, but it had no obvious effect on cytotoxicity. Evidence was obtained which suggested that neutrophils were the main effector cells involved. In subsequent studies with neutrophils [unpublished], it was found using the same occult sera but microfilariae from different dogs, that the cytotoxicity results obtained with heated sera were very variable. Generally, however, the level of cytotoxicity obtained was lower than for non-heated sera. The reasons for this variability need to be further investigated.

In vitro serum-dependent, cell-mediated destruction of microfilariae has now been shown in several host-parasite systems (Mehta et al. 1980, 1981; Haque et al. 1981; Piessens & da Silva 1982), but there has been variation between the systems, in the nature of the effector cells, in the class of antibody involved, and in the requirement for complement in these reactions. The results of the present study with *D. immitis* implied that at least some facets of the cellular adherence, and the cytotoxicity were complement independent, and that the heat stable factor involved may be antibody. Although studies with fractionated sera are required to determine this precisely, this interpretation would be consistent with the observation of Weil et al. (1982) who found that the degree of adherence of canine leucocytes to *D. immitis* microfilariae in the presence of heat-inactivated sera correlated with the titre of IgG antibodies to cuticular microfilarial antigens which they found only in the sera of occult dogs. The present conclusion is also supported by the widespread prevalence of antibodies (IgG class) to microfilarial cuticular antigens in dogs with amicrofilaraemic infections (Dawe, Lewis, McCall, Rawlings & Lindemann 1981; Rawlings et al. 1982).

The finding by Weil et al. (1982) of a heat stable factor in the sera of uninfected dogs which had non-specific opsonic activity resulting in the adhesion of canine leucocytes and platelets to *D. immitis* microfilariae after 1 h incubation was not confirmed in the current study. The reason for this discrepancy in the two studies is not known but it may be related to differences in the methods used for isolating the blood microfilariae. Weil et al. (1982) used hypotonic lysis to remove erythrocytes, whereas in this study microfilariae were separated by filtration.

The present study has also shown that neutrophil preparations of high purity were capable of killing *D. immitis* microfilariae in the presence of appropriate sera. Additional experiments are planned to determine whether or not cytotoxicity in this system was restricted to these cells. It is also important to identify all the cell types capable of participating in the adherence reactions, as there is evidence from the studies of Haque et al. (1981) with *Dipetalonema viteae* microfilariae that some of these adherent cells may facilitate killing by damaging the microfilariae. In the *D. viteae* system described by Haque *et al* (1981) eosinophils were seen to adhere early and cause impairment of microfilarial motility, whereas macrophages which adhered later caused the cytotoxicity.

An interesting, yet unexplained finding of the current study was the heterogeneity observed among the blood microfilariae in their relative susceptibility to serum-dependent, cellular adherence and cytotoxicity. This heterogeneity was most pronounced in cultures containing heat-inactivated occult dog sera. It was unlikely that it was due to a species difference and that the resistant microfilariae belonged to a contaminating *Dipetalonema reconditum* infection as the microfilariae donors used in this study had tested negative for this parasite.

Heterogeneity among microfilariae in other *in vitro* cytotoxicity assays has also been noted. Aiyar, Zaman and Chan (1982) reported that about 5% of *Brugia malayi* microfilariae, although totally covered with cells, remained active and Johnson, Mackenzie, Suswillo & Denham (1981), in studies on *B. pahangi*, observed that there were no cultures in which 100% of the microfilariae were covered with cells. It may be that these differences observed *in vitro* are the basis of a parasite evasion mechanism. Certainly, the operation of such systems *in vivo* may help explain the great variability in the time of onset of the amicrofilaraemic state following either natural or experimental infection. Studies are planned to determine whether the observed microfilarial resistance is due to the masking of antigens by host molecules, or if there are intrinsic differences in the structure of resistant and susceptible microfilariae.

In conclusion, it is suggested that neutrophils and antibodies, probably IgG, were the factors involved in the serum-dependent, cell-mediated destruction of *D. immitis* microfilariae *in vitro*. Since neutrophils and IgG antibodies have been shown to mediate in the *in vitro* killing of the microfilariae of the human filarial species *Wuchereria bancrofti* (Mehta et al. 1981), the present study thus suggests another similarity between *D. immitis* infections in the dog and certain human filarial infections. These similarities have been exploited in recent years and dog dirofilariasis has become a useful experimental model for studying human infections, particularly with regard to pathology (Wong 1974; Castleman & Wong 1982; Weil et al. 1982) and the etiology of adverse reactions to diethylcarbamazine treatment (Palumbo, Desowitz & Perri 1981).

Acknowledgments

I am indebted to Mr R.B. Atwell of the Department of Veterinary Medicine, University of Queensland, for diagnosing the occult infections and collecting the sera, and to Dr C. Bishop (QIMR) for the electron microscopy.

References

Aiyar, S., Zaman, V. and Chan, S.H. 1982. Mechanism of destruction of *Brugia malayi* microfilariae *in vitro*: The role of antibody and leucocytes. *Acta Tropica* **39**:225–36.

Carlisle, C.H. 1980. Canine dirofilariasis: Its radiographic appearance. *Veterinary Radiology* **21**:123–30.

Castleman, W.L. and Wong, M.M. 1982. Light and electron microscope pulmonary lesions associated with retained microfilariae in canine occult dirofilariasis. *Veterinary Pathology* **19**:355–64.

Chalifoux, L. and Hunt, R.D. 1971. Histochemical differentiation of *Dirofilaria immitis* and *Dipetalonema reconditum. Journal of the American Veterinary Medical Association* **158**:601–5.

Dawe, D.L., Lewis, R.W., McCall, J.W., Rawlings, C.A. and Lindemann, B.A. 1981. Evaluation of the indirect fluorescent antibody test in the diagnosis of occult dirofilariasis in dogs. In *Proceedings of the heartworm symposium '80, Dallas, Texas 23–24 February 1980*, ed. G.F. Otto, Kansas: Veterinary Medicine Publishing Company.

Haque, A., Ouassi, A., Joseph, M., Capron, M. and Capron, A. 1981. IgE antibody in eosinophil- and macrophage-mediated *in vitro* killing of *Dipetalonema viteae* microfilariae. *Journal of Immunology* **127**:716–25.

Johnson, P., Mackenzie, C.D., Suswillo, R.R. and Denham, D.A. 1981. Serum-mediated adherence of feline granulocytes to microfilariae of *Brugia pahangi in vitro*: Variations with parasite maturation. *Parasite Immunology* **3**:69–80.

Klei, T.R., Enright, R.M., Blanchard, D.P. and Uhl, S.A. 1982. Effects of presensitization on the development of lymphatic lesions in *Brugia pahangi*-infected jirds. *American Journal of Tropical Medicine and Hygiene* **31**:280–91.

Mehta, K. Sindhu, R.K., Subrahmanyam, D., Hopper, K., Nelson, D.S. and Rao, C.K. 1981. Antibody-dependent cell-mediated effects in bancroftian filariaris. *Immunology* **43**:117–23.

Mehta, K., Sindhu, R.K., Subrahmanyam, D. and Nelson, D.S. 1980. IgE-dependent adherence and cytotoxicity of rat spleen and peritoneal cells to *Litosomoides carinii* microfilariae. *Clinical and Experimental Immunology* **41**:107–14.

Palumbo, N.E., Desowitz, R.S. and Perri, S.F. 1981. Observations on the adverse reaction to diethylcarbamazine in *Dirofilaria immitis* infected dogs. *Tropenmedizin und Parasitologie* **32**:115–18.

Piessens, W.F. and da Silva, W.D. 1982. Complement-mediated adherence of cells to microfilariae of *Brugia malayi. American Journal of Tropical Medicine and Hygiene* **31**:297–301.

Rawlings, C.A., Dawe, D.L., McCall, J.W., Keith, J.C. and Prestwood, A.K. 1982. Four types of occult *Dirofilaria immitis* infection in dogs. *Journal of the American Veterinary Medical Association* **180**:1323–26.

Rzepczyk, C.M. and Bishop, C.J. 1984. Immunological and ultrastructural aspects of the cell-mediated killing of *Dirofilaria immitis* microfilariae. *Parasite Immunology* **6**:443–57.

Subrahmanyam, D., Mehta, K., Nelson, D.S. and Rao, C.K. 1978. Immune reactions in human filariasis. *Journal of Clinical Microbiology* **8**:228–32.

Vadas, M.A., David, J.R., Butterworth, A.E., Pisano, N.T. and Siongok, T.A. 1979. A new method for the purification of human eosinophils and neutrophils, and a comparison of the ability of these cells to damage schistosomula of *S. mansoni. Journal of Immunology* **122**:1228–36.

Weil, G.J., Powers, K.G., Parbuoni, E.L., Line, B.R., Furrow, R.D. and Ottesen, E.A. 1982. *Dirofilaria immitis* VI. Antimicrofilarial immunity in experimental filariasis. *American Journal of Tropical Medicine and Hygiene* **31**:477–85.

Wong, M.M. 1974. Experimental occult dirofilariasis in dogs with reference to immunological responses and its relationship to tropical eosinophilia in man. *Southeast Asian Journal of Tropical Medicine and Public Health* **5**:480–86.

Wong, M.M. and Guest, M.F. 1969. Filarial antibodies and eosinophilia in human subjects in an endemic area. *Transactions of the Royal Society of Tropical Medicine and Hygiene* **63**:796–800.

Wong, M.M., Suter, P.F., Rhode, E.A. and Guest, M.F. 1973. Dirofilariasis without circulating microfilariae: A problem in diagnosis. *Journal of the American Veterinary Medical Association* **163**:133–39.

10 Studies on the Protection of Cattle against *Babesia bovis* Infection

D.F. Mahoney

Antibodies against *Babesia bovis* were first demonstrated serologically in the serum of infected cattle by Mahoney (1962, 1964). This was, however, foreshadowed by Hall (1960, 1963) who found that the calves of infected mothers were more resistant to *B. bovis* infection than the calves of non-infected mothers for at least 2 months after birth. Presumably, resistance was transferred to the calves by antibodies in colostrum. These experiments challenged the concept of premunition (Sergent & Sergent 1956) which had dominated the philosophy of the mechanism of immunity to babesiosis and other similar haemoparasitic diseases for several decades. In brief, the concept stated that antibody formation was minimal, that antibodies were not protective, that immunity depended on the presence of infection and was conferred only by living organisms. As the dogmatic nature of this hypothesis tended to stifle research on the immunology of these diseases it was important to continue investigating the role of antibodies in protection against *B. bovis* in the hope that new areas of knowledge would be revealed. For example, if protection by antibodies was demonstrated, the corresponding target antigen(s) should induce protection, and vaccination against babesiosis with non-living material would be feasible.

The Role of Antibodies in Protection against *B. bovis*

Experimental methods

A passive transfer system to analyse the effect of antibodies on the parasite *in vivo* was developed. Splenectomized calves were used as test animals because parasitaemia after infection by the intravenous (I/V) route developed quantitatively, and provided the basis for a reproducible test system. The experiments consisted firstly of collection and storage of bovine serum from infected and non-infected donors. Then non-infected calves, approximately 100–150 kg in weight, were splenectomized and infected with virulent *B. bovis* parasites. After 4 days, parasites were detectable in thick films of jugular blood, and the calves were then given serum by I/V infusion. The test calves received serum from *B. bovis*-infected donors and the controls received serum from non-infected donors. The doses varied from 2–45 ml/kg body weight. Daily observation of the number of parasites/μl of jugular blood were made for 10–12 days. The results were assessed on the differences in parasitaemia between calves that received *B. bovis* antiserum and their controls. Variations of the experiment were (1) counting parasitaemia every half hour after treatment for kinetic analysis of parasite removal, and (2) using purified gamma globulin extracted from the antiserum to show that the protective activity resided in the antibody fraction.

An *in vitro* test system was also designed. It was first established that the minimum infecting dose of *B. bovis* for calves by the I/V route was between 1 and 10 infected red cells. Doses of 10, 100 and 1000 infected red cells were then incubated at 37°C for 1 h in 10 ml of antiserum and inoculated along with appropriate control samples into calves for tests of infectivity. In some experiments fresh bovine complement was added to the antiserum and in others the bovine erythrocytes in the inoculum were lysed with sheep antiserum against cattle red cells and guinea pig complement before exposure to the *B. bovis* antiserum.

The effect of antiserum on parasitaemia

Immediately after treatment of an infected calf with antiserum, the parasites commenced to disappear from the blood, and fell below the detectable level within 24–48 h. The rate and degree of the reaction depended on the dose and type of antiserum administered. Serum from donors that had been infected several times at intervals of 3–6 months (i.e. hyperimmune serum) was effective in low doses (e.g. 2.2 ml/kg live weight) (fig.10.1a), whereas antiserum taken soon after recovery from initial infection was less effective and higher doses (22 ml/kg) were required to demonstrate protection (fig.10.1b).

Similarly, gamma-globulin solutions purified from hyperimmune serum were also effective and kinetic studies on the disappearance of the parasites showed that the antibodies attacked the infected cells as well as emerging organisms (Mahoney, Kerr, Goodger & Wright 1979). Thus the target antigens were located on the surface of infected red cells and on the parasites. Antibody, however, had no effect on the infectivity of parasitized red cells *in vitro*, and clearly required the environment of the living host to function (Mahoney 1972). The antibody probably acted as an opsonin, the organisms being killed by phagocytes. Although no definitive work has been performed on the phagocytosis of *B. bovis*, Rogers (1974) correlated opsonic activity of antiserum to *B. rodhaini* with its ability to protect rats in passive transfer tests. The suitability of splenectomized calves for passive transfer tests was not regarded as contrary to this hypothesis. Because of its size, the liver is a more important phagocytic organ than the spleen (Taliaferro 1956) and splenectomized animals were thus expected to retain a significant part of their phagocytic capability. In addition, both splenectomized and intact animals were used during preliminary work on the development of the

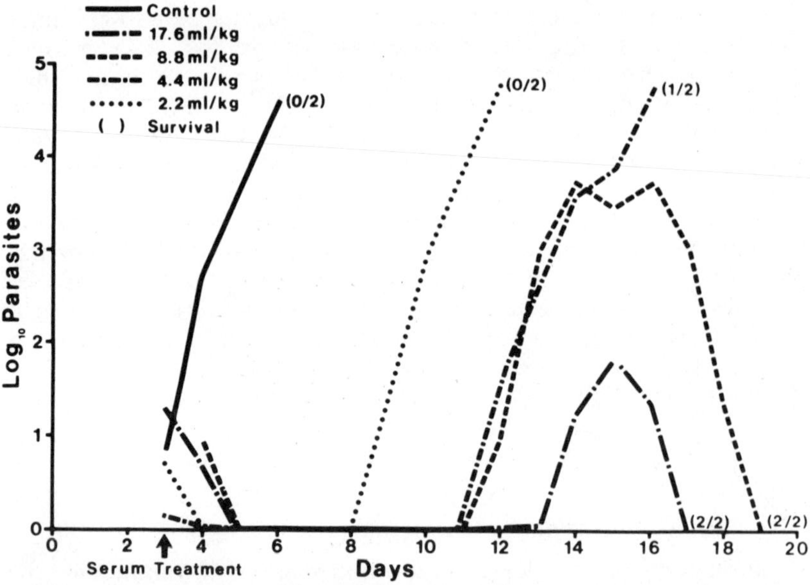

Figure 10.1a Parasitaemia and survival of *B. bovis*-infected calves after treatment with 2.2, 4.4, 8.8 and 17.6 ml of hyper-immune antiserum per kg body weight. (From Mahoney, D.F., Kerr, J.D., Goodger, B.V. & Wright, I.G. 1979. *International Journal for Parasitology* **9**:297–306.)

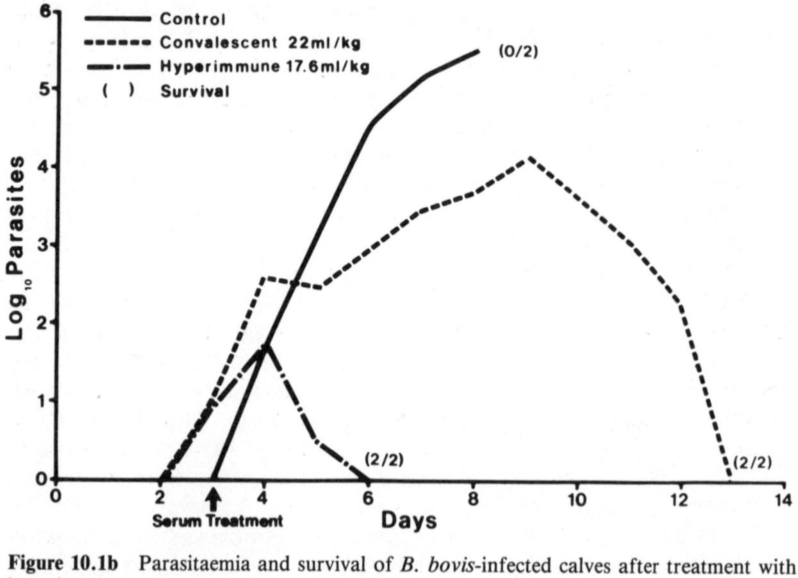

Figure 10.1b Parasitaemia and survival of *B. bovis*-infected calves after treatment with hyperimmune serum and serum taken from donors 2 weeks after recovery from initial infection (convalescent).

passive transfer system (Mahoney 1967a). No qualitative differences were observed in the reactions that occurred in the two types of animal, but splenectomized groups provided a test of higher sensitivity.

Relapse of parasitaemia was a feature of the passive transfer experiments when hyperimmune serum was given in low dosage (2–4 ml/kg). Higher doses (10–20 ml/kg) caused a progressive diminution in the height and severity of such relapses and most calves that received over 15 ml/kg did not show them (fig.10.1a). The relapses were simply the result of the quantitative relationship between factors such as the number of parasites in the blood at the time of serum administration, the initial concentration of antibody and its rate of wastage on antigen-antibody reactions and normal catabolism. They were not caused by antigenic variation of the parasite population induced by the action of protective antibody because the relapse parasites were susceptible to the same serum on re-exposure (Mahoney et al. 1979).

The relationships between antigenic and strain variation and the activity of protective antibodies

A population of parasites isolated at one time from the field was designated as a strain and named after the locality from which it came. Each isolate (strain) was maintained as a separate population in the laboratory by cryopreservation in liquid nitrogen. Three strains, viz., Lismore (L), Samford (S) and Helidon (H), were used to infect splenectomized calves, and hyperimmune serum against two, Gayndah (G) and L, was used for treatment of the infected animals. All sera were tested for activity against the homologous parasites before use in the experiments. Two experiements were performed and the dose of antiserum administered was 17.6 ml/kg. In the first experiment, 4 animals were infected with the L strain and treated with G strain antiserum. In the second, groups of 2 animals were infected with H and S strains respectively and treated with L strain antiserum. In one additional experiment 2 animals infected with L strain were given G strain antiserum at a dose of 44 ml/kg. In all cross-protection experiments, the antiserum treatment had no effect on the parasites of the heterologous strains.

Another type of variation in the antigenic character of *B. bovis* was described by Curnow (1968, 1973). This occurred with strains during the period of chronic infection in individual animals. Mahoney (1962) showed that chronic infection with *B. bovis* lasted for several years and was characterized by sub-clinical relapses of detectable parasitaemia in peripheral blood at irregular intervals. Curnow showed that each such relapse was associated with a different agglutinogen on the surface of the infected red blood cells. His hypothesis was that the variable agglutinogens were targets for protective antibodies and that the variation of specificity was the parasite's strategy to avoid the host's immune response and maintain chronic infection.

The relationship between the protective strain-specific immunity observed in passive transfer experiments and the variant-specific serological activity during chronic infection had to be determined because of implications for the development of a killed vaccine. If the two phenomena were associated, the likelihood of developing a killed vaccine of relatively simple antigenic composition seemed remote. Experiments, however, suggested that the two immunological events were manifestations of different antigen-antibody reactions (Mahoney et al. 1979). In passive transfer tests, there was no difference in the specificity of protective antibodies taken from a chronically infected animal at different times in relation to protection against the relapse sub-strains from the same animal. Thus protective antibodies generated by one variant population from within a strain were effective against all other variant populations produced by that strain.

The Use of Antigen to Induce Protection

Experimental methods

The basic requirement for studies on the antigens of *B. bovis* was to obtain a concentrated source of the organism. This requirement was met by the separation of infected from non-infected cells in blood (Mahoney 1967b). The technique was based on the observation that the infected erythrocytes were less susceptible to hypotonic lysis than the uninfected cells. It was therefore possible to select a concentration of salt solution that lysed all uninfected erythrocytes leaving the infected ones intact to be recovered by differential centrifugation. The method was rapid, reproducible and applicable on a preparative scale. It was used to produce suspensions of 95%–100% infected erythrocytes from blood with the parasitaemia in the range of 5% to 15%. Extracts of crude antigen were prepared from these infected cell suspensions by sonic disintegration for 2–4 min at maximum power of the instrument and then centrifugation at 145,000 *g* for 60 min at 4°C. The supernatant fluid was the crude soluble antigen and it was fractionated by precipitation with protamine sulphate, by gel filtration, and by affinity chromatography using antibodies from immune cattle (fig.10.2).

Tests for the immunogenic activity of antigens consisted of the subcutaneous inoculation of the antigen as a water-in-oil emulsion with Freund's Complete Adjuvant (FCA) into two-year-old steers. In early work 3 inoculations, 2 weeks apart, were used but in later studies the number of inoculations was reduced to 2 and to 1. Two to 4 weeks after immunization, the cattle were challenged by I/V inoculation with a different strain of *B. bovis* from the one used to prepare the antigen. The immune response was then assessed by comparing daily rectal temperatures, levels of parasitaemia and falls in packed cell volume in

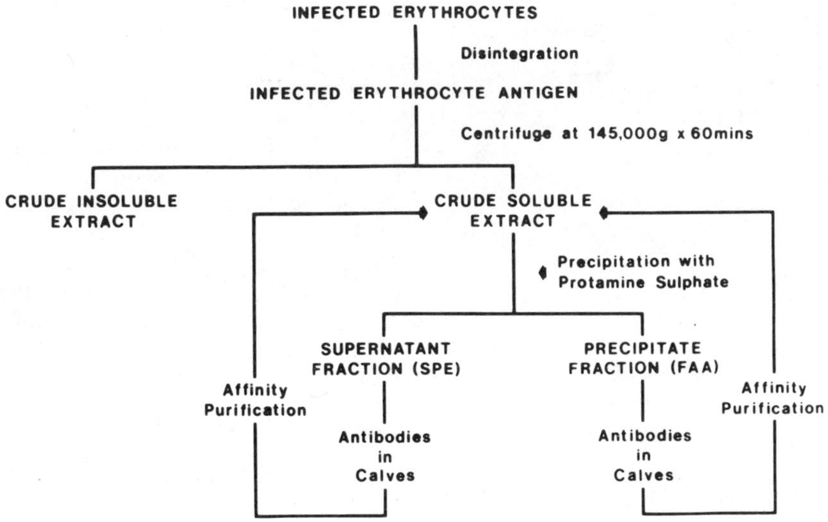

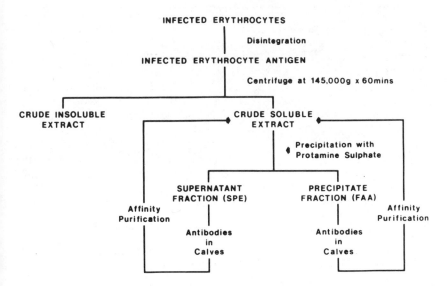

Figure 10.2 Flow diagram of the fractionation procedure for bovine erythrocytes infected with *B. bovis*. (From Mahoney, D.F., Wright, I.G. & Goodger, B.V. 1984. *Preventive Veterinary Medicine* 2:401–8.)

vaccinated and control groups. The latter were sham-immunized with FCA alone.

Protection by crude antigen

Mahoney and Wright (1976) showed that immunization with crude antigen prepared from infected erythrocytes protected cattle against heterologous *B. bovis* infection as effectively as immunity associated with active infection (fig.10.3a). This material was a mixture of all particulate and soluble components contained by infected erythrocytes. It was fractionated into soluble and insoluble components by ultracentrifugation and both fractions induced similar protection (fig.10.3b), demonstrating that protective antigen was released into solution.

Fractionation of the crude soluble antigen

The fractionation procedure was based on studies by Goodger (1971, 1973, 1976) who showed that the antigens extracted from the *B. bovis*-infected erythrocyte contained several different specificities and he classified them into three groups. The first group was composed of two

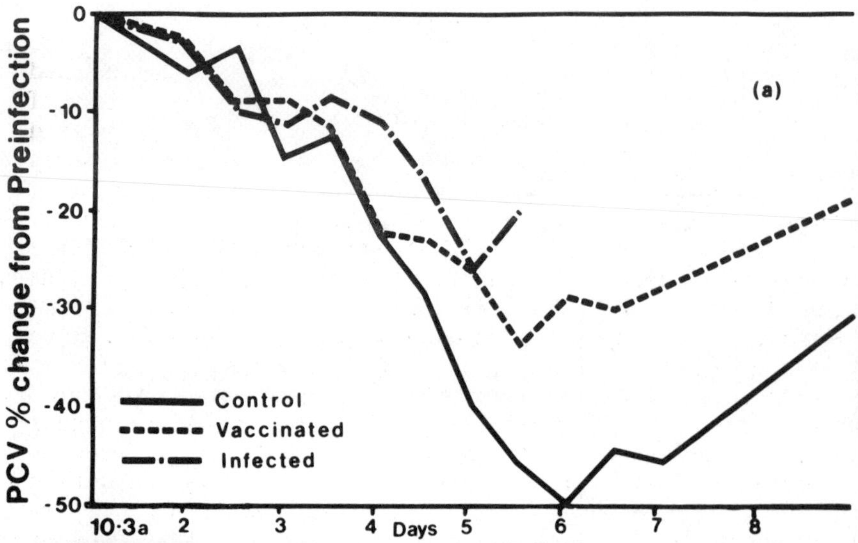

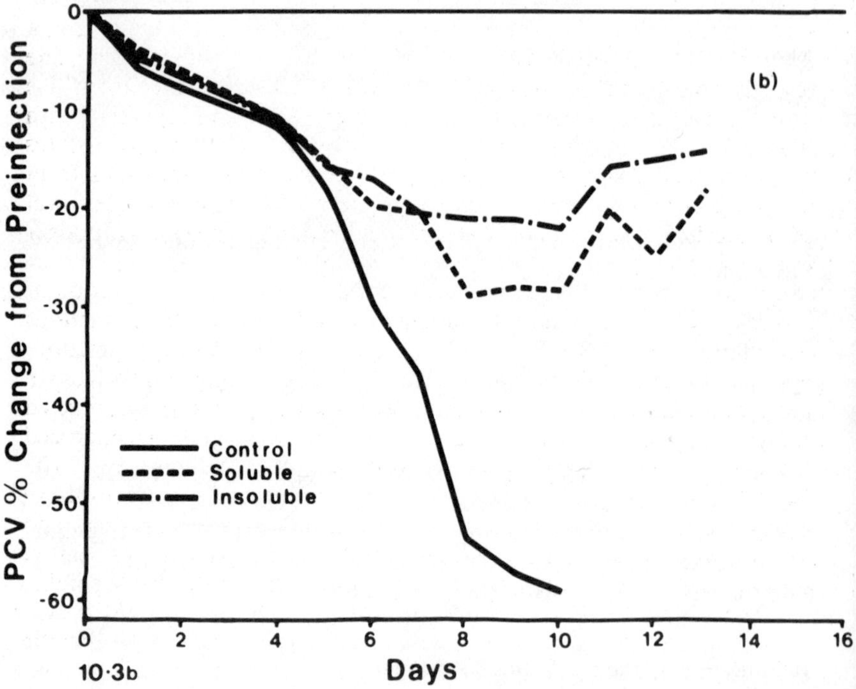

Figure 10.3a Comparison of the changes in packed cell volume (PCV) in cattle vaccinated with crude *B. bovis* antigen and in cattle immunized by *B. bovis* infection, after challenge with virulent heterologous *B. bovis* organisms. (From Mahoney, D.F. & Wright, I.G. 1976. *Veterinary Parasitology* **2**:273–82.)

Figure 10.3b The changes in PCV in cattle vaccinated with soluble and insoluble fractions of crude *B. bovis* antigen and challenged with an heterologous strain of the organism. (From Mahoney, D.F., Wright, I.G. & Goodger, B.V. 1984. *Preventive Veterinary Medicine* **2**:401–8.)

antigens associated with the stroma of the erythrocyte. These were responsible for staining of the membrane by fluorescein-labelled antibodies (C.G. Ludford, unpublished Ph.D. thesis, University of Queensland, 1967). One of the antigens was located as a dense band in or under the cell membrane and the other had a granular distribution throughout the stroma. Characterization of the stromal antigens showed that they were basically fibrinogen molecules, altered by conjugation with a number of peptides, some of which were of babesial and others of host origin (Goodger, Wright, Mahoney & McKenna 1980). Separation of the fibrinogen-associated antigen complexes was achieved by methods that specifically precipitated fibrinogen (Goodger 1976). The second group was composed of antigens located on the parasite. They differed in specificity from those on the stroma and membrane, but little was known of their physical and chemical properties. Another antigen was found in the cytoplasm of the infected erythrocyte and extracted from the haemoglobin solution obtained after lysis of the erythrocytes in distilled water. The fibrinogen-associated antigens were precipitated from the crude soluble extract with protamine sulphate and this step effectively separated those antigens that were associated with the erythrocyte stroma from those located on the parasite and in the erythrocyte cytoplasm. The response in calves inoculated with each fraction seemed to confirm this broad separation because antibodies from those immunized with fibrinogen-associated antigen stained the stroma of infected erythrocytes in the indirect fluorescent antibody (IFA) test and the antibodies from the group immunized with the antigen(s) left in solution after the removal of the precipitate stained only the parasites (Mahoney, Wright & Goodger 1981).

Both fractions protected cattle against challenge. One interpretation of this result was that an antigenic component, common to both fractions but not distinguisable by IFA test was responsible for protection. However, there could be more than one target antigen involved in protection against *B. bovis* and these could be located on both the infected red cells and the parasites. For example, Mahoney et al. (1979) concluded that the protective antibodies attacked targets at both sites. The immunoglobulins from the serum of calves inoculated with each of the above fractions (i.e. precipitate and supernatant) were purified and coupled to CNBr Sepharose 4B columns for affinity chromatography. These columns were then used to extract antigen from the crude soluble material (fig.10.2). The concentration of protein in the extracts was approximately 200 μg/ml and contained in addition to *B. bovis*-specific material, components from normal bovine erythrocytes owing to non-specific absorption of such material on the Sepharose columns. Each fraction contained three antigenic components from *B. bovis*, but there were no reactions of identity or even partial-identity between the two groups in immunodiffusion tests. Cattle were immunized with each fraction by single injection. Two dose levels, 100 μg and 500 μg of protein, were used. With the antigens from the supernatant fraction (SPE), the dose of 100 μg of protein gave protection (fig.10.4a) but the higher dose was less protective. With the antigens isolated from crude soluble material by antibodies to the precipitate (FAA), the 500 μg dose gave more effective protection (fig.10.4b).

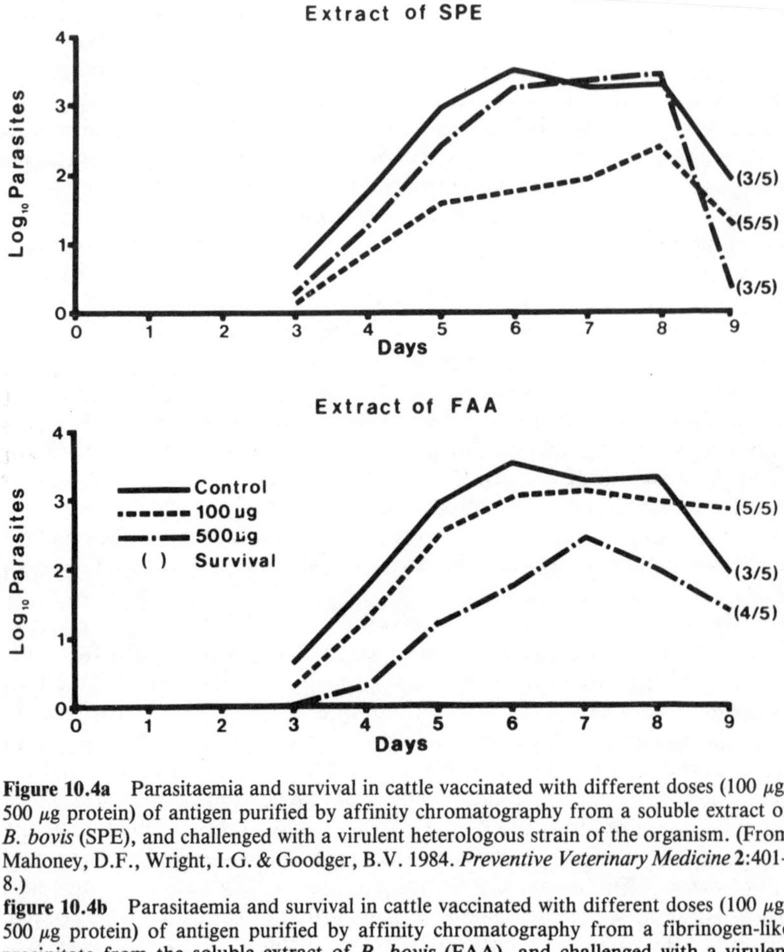

Figure 10.4a Parasitaemia and survival in cattle vaccinated with different doses (100 μg, 500 μg protein) of antigen purified by affinity chromatography from a soluble extract of *B. bovis* (SPE), and challenged with a virulent heterologous strain of the organism. (From Mahoney, D.F., Wright, I.G. & Goodger, B.V. 1984. *Preventive Veterinary Medicine* **2**:401–8.)
figure 10.4b Parasitaemia and survival in cattle vaccinated with different doses (100 μg, 500 μg protein) of antigen purified by affinity chromatography from a fibrinogen-like precipitate from the soluble extract of *B. bovis* (FAA), and challenged with a virulent heterologous strain of the organism. (From Mahoney, D.F., Wright, I.G. & Goodger, B.V. 1984. *Preventive Veterinary Medicine* **2**:401–8.)

Discussion

Aspects of the data on passive transfer of immunity and on immunization with antigen must be reconciled. On one hand, the inoculation of the host with extracts of parasitized erythrocytes from one strain protected against infection with another strain. In contrast, the antibodies that were present in the serum of animals immunized with one strain had no effect on the parasites of another strain. The determinants of the protective antigens of each strain were apparently different. Observations on the immune response after heterologous challenge showed that animals immunized either by antigen or by active infection reacted in a manner similar to the susceptible controls for approximately 6 days after infection and then the protective reaction was manifest. The event that preceded recovery was probably the production of protective antibody specific for the challenge strain. To accomplish this in sufficient time for protection against disease, the animals must have been primed for a secondary reaction against the unrelated antigenic determinant(s) of the new strain. Thus there must be a hidden relationship between the apparently different protective antigens of *B. bovis* strains. This could involve that portion of the antigenic molecules not concerned directly in the reaction with protective antibody, i.e., the carrier moiety. The role of the carrier in priming the host for a secondary response to hapten-carrier conjugates is well recognised (Mitchison 1971a, 1971b) and a model for carrier-specific enhancement of the immune response to related determinants which appears to be relevant to the *B. bovis* situation has been studied in mice (Terres, Habicht & Stoner 1974). This question can only be answered by the isolation of the protective antigens from cloned strains of *B. bovis* and a comparison of amino-acid sequences.

The feasibility of vaccination against *B. bovis* with a killed antigen appears to have been established. Antigenic variation of the parasite does not circumvent the host's protective response and the nature of the antigenic differences between strains, although poorly understood at the molecular level, is obviously not the type that prevents cross-protection in the field.

A successful vaccine derived from *B. bovis* infected erythrocytes should have the following characteristics: (1) prevention of clinical disease; (2) efficacy against all immunological strains of the parasite; (3) one or two inoculations should induce protection; (4) protection should be induced for a minimum of 6 months; (5) no concurrent immunization against host blood group antigens; (6) availability in large quantities; and (7) stability on storage. The criteria (1), (2) and (3) above have been fulfilled. The duration of protection after immunization with antigen has not been determined. However, it has been found that immunity to reinfection remained for at least 6 months after termination of infection with chemotherapy (Callow, McGregor, Parker & Dalgliesh 1974) and this suggested that the longevity of immunity induced by antigen inoculation would be of similar duration.

The production of antigen is the most important unknown factor in relation to vaccine development. The growth of parasites in calves would not provide enough antigenic material for even a small vaccination program. A culture system for *B. bovis* in bovine blood has been developed (Levy & Ristic 1980), but as erythrocyte material is still present the difficulty of separating it from the parasite antigen remains. The alternative is to isolate and characterize the protective antigen(s) from the parasite using monoclonal antibodies, and from this step proceed to the production of the antigen(s) by recombinant DNA methods. This approach should solve both erythrocytic contamination and production problems in the same operation, and result in a relatively cheap and stable vaccine.

Acknowledgments

The author is grateful to Elsevier Science Publishers B.V. and the Australian Society for Parasitology for permission to reproduce previously published material.

References

Callow, L.L., McGregor, W., Parker, R.J. and Dalgliesh, R.J. 1974. The immunity of cattle to *Babesia argentina* after drug sterilisation of infections of varying duration. *Australian Veterinary Journal* **50**:6–11.

Curnow, J.A. 1968. *In vitro* agglutination of bovine erythrocytes infected with *Babesia argentina*. *Nature (London)* **217**:267–68.

Curnow, J.A. 1973. Studies on antigenic changes and strain differences in *Babesia argentina* infections. *Australian Veterinary Journal* **49**:279–83.

Goodger, B.V. 1971. Preparation and preliminary assessment of purified antigens in the passive haemagglutination test for bovine babesiosis. *Australian Veterinary Journal* **47**:251–56.

Goodger, B.V. 1973. *Babesia argentina*: Intraerythrocytic location of babesial antigen extracted from parasite suspensions. *International Journal for Parasitology* **3**:387–91.

Goodger, B.V. 1976. *Babesia argentina*: Studies on the nature of an antigen associated with infection. *International Journal for Parasitology* **6**:213–16.

Goodger, B.V., Wright, I.G., Mahoney,D.F. and McKenna, R.V. 1980. *Babesia bovis (argentina)*: Studies on the composition and location of antigen associated with infected erythrocytes. *International Journal for Parasitology* **10**:33–36.

Hall, W.T.K. 1960. The immunity of calves to *Babesia argentina* infection. *Australian Veterinary Journal* **36**:361–66.

Hall, W.T.K. 1963. The immunity of calves to tick-transmitted *Babesia argentina* infection. *Austrlian Veterinary Journal* **39**:386–89.

Levy, M.G. and Ristic, M. 1980. *Babesia bovis*: Continuous cultivation in a microaerophilous stationary phase culture. *Science* **207**:1218–20.

Mahoney, D.F. 1962. Bovine babesiosis: Diagnosis of infection by a complement fixation test. *Australian Veterinary Journal* **38**:48–52.

Mahoney, D.F. 1964. Bovine babesiosis: An assessment of the significance of complement fixing antibody based upon experimental infection. *Australian Veterinary Journal* **40**:369–75.

Mahoney, D.F. 1967a. Bovine babesiosis: The passive immunization of calves against *Babesia argentina* with special reference to the role of complement fixing antibodies. *Experimental Parasitology* **20**:119–24.

Mahoney, D.F. 1967b. Bovine babesiosis: Preparation and assessment of complement fixing antigens. *Experimental Parasitology* **20**:232–41.

Mahoney, D.F. 1972. Immune response to hemoprotozoa. II. *Babesia* spp. In *Immunity to animal parasites*, ed. E.J.L. Soulsby, 301–41. New York: Academic Press.

Mahoney, D.F., Kerr, J.D., Goodger, B.V. and Wright, I.G. 1979. The immune response of cattle to *Babesia bovis* (syn. *B. argentina*). Studies on the nature and specificity of protection. *International Journal for Parasitology* **9**:297–306.

Mahoney, D.F. and Wright, I.G. 1976. *Babesia argentina*: Immunization of cattle with a killed antigen against infection with a heterologous strain. *Veterinary Parasitology* **2**:273–82.

Mahoney, D.F., Wright, I.G. and Goodger, B.V. 1981. Bovine babesiosis: The immunization of cattle with fractions of erythrocytes infected with *Babesia bovis* (syn. *B. argentina*). *Veterinary Immunology and Immunopathology* **2**:145–56.

Mitchison, N.A. 1971a. The carrier effect in the secondary response to hapten-protein conjugates. I. Measurement of the effect with transferred cells and objections to the local environment hypothesis. *European Journal of Immunology* **1**:10–17.

Mitchison, N.A. 1971b. The carrier effect in the secondary response to hapten-protein conjugates. II. Cellular cooperation. *European Journal of Immunology* **1**:18–27.

Rogers, R.J. 1974. Serum opsonins and the passive transfer of protection in *Babesia rodhaini* infections of rats. *International Journal for Parasitology* **4**: 197–201.

Sergent, E. and Sergent Et. 1956. History of the concept of "relative immunity" or "premunition" correlated to latent infection. *Indian Journal of Malariology* **10**:53–80.

Taliaferro, W.H. 1956. Functions of the spleen in immunity. *American Journal of Tropical Medicine and Hygiene* **5**:391–410.

Terres, G., Habicht, G.S. and Stoner, R.D. 1974. Carrier-specific enhancement of the immune response using antigen-antibody complexes. *Journal of Immunology* **112**:804–11.

Part Four

Host-Parasite Interactions

11 Larval Taeniid Cestodes — Models for Research on Host-Parasite Interactions

M.D. Rickard

Introduction

Larval taeniid cestodes are important parasites because of the adverse effects they have on human health as well as the economic losses occasioned by the condemnation of infected carcases of domesticated animals. The past decade has seen a surge of research interest in these parasites not only because of their intrinsic importance, but also because of the many advantages they offer as models for studying important aspects of host-parasite relationships. Host modulation of infection with larval cestodes plays a central role in regulating their natural transmission (Gemmell 1976; Gemmell & Johnstone 1977) and in few parasitic associations is this phenomenon so clear cut. In addition, many of the more serious consequences of infection with these parasites result from activation of the cellular and humoral defence mechanisms of the host.

The purpose of this paper is to outline briefly some aspects of the interaction between these parasites and their hosts in order to highlight their advantages as models for fundamental and applied research on this topic.

The Parasites

Larval taeniid cestodes of major importance to human health include *Echinococcus* spp., especially *E. granulosus* and *E. multilocularis*, the causative agents of hydatid disease, and *Taenia solium*, the aetiological agent of cysticercosis in humans. *T. ovis* and *T. saginata* cause economic losses in sheep and cattle respectively because the unsightly lesions they produce in the musculature make it unfit for human consumption. Infection of sheep and cattle with *E. granulosus* or *T. hydatigena* results in the condemnation of offal, and *T. multiceps* larvae can cause serious CNS disturbances in sheep and goats.

Infection with these parasites is initiated by accidental ingestion of eggs released from gravid proglottides in the faeces of definitive hosts carrying the adult cestode. The oncosphere larva is surrounded by a thick embryophore made up of small blocks of keratinaceous material. Under the influence of gastric and intestinal secretions, the cement substance binding the blocks swells, disrupts the embryophore and releases the oncosphere enclosed in its oncospheral membrane. The oncosphere becomes active under the influence of bile and other gut secretions and tears its way out of the enclosing membrane by means of its three pairs of hooks. It then penetrates the intestinal epithelium, enters the lamina propria and travels via lymphatic or blood vessels to the site of election. Some developing larvae undergo substantial tissue migration before finally maturing into the sedentary, cystic metacestode stage which is surrounded by a capsule of host fibrous tissue.

Some of the more important cestodes present problems for experimental work. The eggs of *E. granulosus* and *T. solium* are infective for humans and therefore difficult to handle and *T. solium* and *T. saginata* eggs are difficult to obtain in large quantities because they can only be collected from adult worms in infected humans. The life cycle of *T. ovis* is not easily maintained in the laboratory because it is usually difficult to obtain adequate numbers of viable cysticerci with which to infect dogs unless experimentally infected lambs are immunosuppressed using corticosteroids; cysticerci obtained from abattoir material are mostly non-viable. *T. hydatigena* is usually the easiest of the taeniids infecting domestic animals to work with experimentally. Viable cysticerci for infecting dogs are common in abattoir material, or if not, sheep are easily infected experimentally. Dogs are readily infected and 5 mature tapeworms from a dog may yield up to 5×10^6 eggs from their terminal 15–20 proglottides. The predilection sites for cysticerci of this parasite in sheep are the liver and peritoneal cavity which makes them readily accessible for counting or other manipulations.

In addition to the species in larger domesticated animals, several laboratory models are available for study. The cat-mouse (rat) tapeworm *T. taeniaeformis* and the dog-rabbit worm *T. pisiformis* are very useful. These are natural host-parasite associations and the parasite and host behave in much the same way as do the important species. *T. taeniaeformis* in the mouse is particularly useful because inbred strains of mice with uniform receptivity to infection are readily available and the immune response of the mouse is so well characterized. One unfortunate by-product of the convenience and lesser cost of maintaining these laboratory models is that there is a dearth of fundamental information available on host responses to infection with the more important species which are more difficult to work with.

Taeniid eggs are easy to handle and store (Coman & Rickard 1975), and although best infections are obtained with fresh material, eggs will remain infective for 1–2 months if stored aseptically at 4°C. They do not require any period of maturation or development outside of the definitive host.

Care has to be taken in comparing experiments with parasite material obtained from different sources. There is increasing evidence of "strain" variations in infectivity amongst cestodes. This has been documented with *E. granulosus* by Thompson (1978, 1979), Thompson and Kumaratilake

(1982); and *T. taeniaeformis* by Heath and Elsden-Dew (1972), Ambu and Kwa (1980), Brandt and Sewell (1981) and Williams, Shearer and Ravitch (1981). Strains of *T. taeniaeformis* have shown markedly different infectivities for rats and mice (Ambu & Kwa 1980; Brandt & Sewell 1981) as well as varying infectivity for different age groups within a single species (Brandt & Sewell 1981).

Some parasites can be maintained by serial passage of the metacestode stage thus eliminating the requirement for keeping definitive host animals. *E. granulosus* develops readily in mice, *E. multilocularis* in several rodents including mice, and *T. crassiceps* in laboratory mice and rats. These parasites have been widely used in laboratory studies, but some care must be applied when interpreting experiments. Continual passage without recourse to egg infection can result in biological changes in the parasite. In addition, the response of the host to implanted metacestodes is almost certainly different in many respects to that following egg infection and exposure of the host to all developmental phases. Nevertheless, the polyembryonic cestodes are useful for studying mechanisms which allow the prolonged survival of metacestode stages in immune hosts, as well as for developing methods for immunotherapy or chemotherapy.

Genetic Variations in Resistance

Detailed information on genetic variation in resistance to infection with larval taeniid cestodes is available in recent reviews (Wakelin 1976; Rickard & Williams 1982). There are marked variations in resistance to both primary and secondary infections with larval cestodes due to factors such as strain, sex, age and physiological status of the host.

In outbred populations of domesticated animals it is extremely difficult to accurately quantify differences in innate resistance, but the marked variations in individual "takes" of infection following a uniform dose of eggs underline their importance. Breed differences in resistance have not been reported, but sex differences in susceptibility to naturally acquired infection with *T. saginata* in cattle have been described (Ginsberg, Cammeron, Goddard & Grieve 1956; Froyd 1960). The practical importance of sex differences in susceptibility to infection has been highlighted by the finding that male cattle respond poorly to artificial immunization against *T. saginata* infection by comparison with females (Rickard, Brumley & Anderson 1982).

Genetic variations in susceptibility to initial infection have been documented in different inbred strains of laboratory rodents, e.g. *T. crassiceps* in rats (Chernin 1981) and mice (Siebert & Good 1980); *T. taeniaeformis* in rats (Curtis, Dunning & Bullock 1933; Olivier 1962; Williams et al. 1981) and mice (Dow & Jarrett 1960; Olivier 1962; Orihara 1962; Mitchell, Goding & Rickard 1977); *E. multilocularis* in mice (Yamashita, Ohbayashi, Kitamura, Suzuki & Okugi 1958; Lubinsky

1964). These variations in resistance are expressed by reduced establishment of larvae as well as limited survival of those that commence development.

Variations in host physiological and anatomical factors undoubtedly play a significant part in innate resistance to infection. For example, a high proportion of cestode oncospheres fail to penetrate the gut of older animals (Heath 1971; Turner & McKeever 1976) and it has been suggested (Musoke, Williams, Leid & Williams 1975) that rats less than 2–3 months old are resistant to *T. taeniaeformis* infection because they lack proteolytic enzymes in their intestine. However, a number of host defence mechanisms have also been implicated in natural resistance.

The responsiveness of non-specific cellular defences may account in part for variations in innate resistance. Enhanced cellular responses were described in older mice as early as 24–48 h post-infection (Turner & McKeever 1976). Furthermore, *E. granulosus* protoscolices stimulate blastic proliferation of normal, unprimed murine T lymphocytes (Dixon, Jenkins, Allan & Connor 1978; Dixon, Jenkins & Allan 1982). BCG is known to enhance non-specific cellular defences and injection with BCG stimulates protection against infection with *E. multilocularis* protoscolices in cotton rats (Rau & Tanner 1975; Reuben, Tanner & Rau 1978), *E. granulosus* protoscolices in gerbils (Thompson 1976) and *T. taeniaeformis* eggs in mice (Thompson, Penhale, White & Pass 1982). In the latter case, both establishment and subsequent survival of larvae were reduced in mice injected with BCG.

The effect of BCG may be due not only to enhancement of non-specific cellular defence mechanisms but also to promoting a more rapid specific immune response to the parasite. The rate at which antibody responses occur has been used to explain the variations in innate resistance to *T. taeniaeformis* infection in mice (Mitchell, Rajasekariah & Rickard 1980). The survival of taeniid larvae depends initially upon a race between the defence mechanisms of the parasite and those of the host and these workers (Mitchell et al. 1980) found that innately more resistant strains of mice produced antibody more rapidly following infection than did the susceptible strains.

BALB/c and BDF1 mice encapsulate *T. crassiceps* larvae more rapidly than C3H mice (Siebert & Good 1980) and it has been shown (Pollaco, Nicholas, Mitchell & Stewart 1978) that the encapsulating response to *Mesocestoides corti* infection in mice is T cell dependent. Variations in complement activity in different strains of mice may also play a part in determining innate resistance (Mitchell, Goding & Rickard 1977) because complement-fixing antibody has a central role in acquired resistance to larval taeniids.

An understanding of the genetic basis for variations in innate resistance is of utmost practical significance. If genetic markers for resistance can be defined it may be possible to selectively breed resistant animals. Also, variations in innate resistance may determine the success or failure of vaccination programmes. It has already been pointed out that sex of the host influences the efficacy of immunization (Rickard et al. 1982) and the results of experiments on *T. taeniaeformis* infection in mice suggest that there is also an interaction between strain of mouse and efficiency of vaccination (Rickard, Rajasekariah & Mitchell 1981). Very limited progress has thus far been made on the analysis of genetic control of resistance to these parasites (Lubinsky 1964; Mitchell et al. 1980). Larval taeniids provide excellent models for studies on the interactions between artificial immunization and genetic resistance, and such investigations are of high priority.

Immunity

There are several recent reviews concerning immunity to larval taeniid cestodes (Flisser, Pérez-Montfort & Larralde 1979; Williams 1979; Rickard & Williams 1982). Immunity directed against the establishment of egg-induced infection is striking and there has been substantial research on this topic. However, many issues remain unresolved. Clearly, complement-fixing serum antibody plays a major role in defence and extensive analyses of antibody subclasses responsible for immunity to infection against *T. taeniaeformis* infection in mice and rats have been carried out (see Williams 1979; Rickard & Williams 1982). In addition to serum antibody, intestinal and colostral IgA antibody are also highly effective in passive immunization (Lloyd & Soulsby 1978). Although it is known that the oncosphere and early post-oncospheral stages are highly susceptible to antibody-mediated attack, definitive information on the mechanism of antibody action is not available. Virtually no information is available on the classes of antibody responsible for protection in domesticated animals.

The role of antibody-dependent cell-mediated cytotoxicity in preventing establishment of larval taeniids is much less well documented. Extensive studies have shown the importance of eosinophils (see Butterworth, Vadis & David 1980), macrophages (Capron, Dessaint, Capron & Bazin 1975) and mast cells (Capron, Capron, Torpier, Bazin, Bout & Joseph 1978) in the destruction of schistosome stages *in vitro*, although the relevance of these findings to events *in vivo* is not yet established. Similar detailed studies on larval cestodes are lacking, although there is circumstantial evidence for the participation of several cell types in the immune response.

IgE antibody responses occur in *T. taeniaeformis* infection in rats and *T. pisiformis* in rabbits (Leid & Williams 1974, 1975). Although this antibody is not essential for the passive transfer of resistance, immune sera containing IgE enhance the level of protection achieved. The rapid death of oncospheres incubated in serum *in vitro* (Silverman 1955; Rickard & Outteridge 1974) contrasts with the observation by Musoke and Williams (1975) that *T. taeniaeformis* larvae were still alive in the liver 24 h after infection of mice passively immunized with sera containing only IgG_{2a} (Musoke & Williams 1975). When sera containing IgE in addition to IgG_{2a} were used, larvae were killed within 12 h (Musoke, Williams & Leid 1978). Mast cells accumulate around developing larvae (Coleman & De Salva 1963; Singh & Rao 1967; Varute 1971; Siebert, Good & Simmons 1979; M. Lindsay, unpublished M.S. thesis, Michigan State University, 1981) and it has been suggested that IgE may interact with antigen and cause degranulation of mast cells with the consequent increase in vascular permeability allowing easier access of IgG antibodies to the site of invading or developing larvae. Mast cells may play a more direct role in defence because intestinal penetration of *T. taeniaeformis* oncospheres is reduced following direct injection of peritoneal anaphylactic diffusate into the intestinal lumen of mice (Musoke et al. 1978).

Eosinophils accumulate around cestode larvae (see Ansari & Williams 1976) and secondary eosinophil responses to *T. taeniaeformis* infection in rats have been described (Ansari & Williams 1976). This eosinophil response is antibody-dependent (Heath & Pavloff 1975; Ansari, Williams & Musoke 1976), but there is as yet no evidence to suggest that eosinophils in any way harm oncospheres or post-oncospheral stages *in vivo* or *in vitro*. Similarly there is no evidence for a role of macrophages or neutrophils, although these cells also accumulate at the site of dying oncospheres (Furukawa, Niwa & Miyazato 1981).

Obviously there is a vast amount of information yet to be obtained concerning the role of antibody and antibody-mediated cellular effects in immunity to larval cestodes. Improvements in methods for the *in vitro* cultivation of taeniid cestode larvae (Heath 1973a; Heath & Lawrence 1976; Lawrence, Heath, Parmeter & Osborn 1980) may provide a useful tool to examine these immune mechanisms; a study of the effect of antibody on *in vitro* development of *E. granulosus* larvae has been reported (Heath & Lawrence 1981).

Lymphocyte-mediated cytotoxicity directed against early post-oncospheral stages has not been documented. Indirect observations concerning the effects of thymectomy, anti-lymphocyte serum or adoptive transfer of cells have not been particularly helpful because they have not excluded the role of T-cell-dependent antibody-mediated effects (Okamoto 1968; Németh 1970; Okamoto & Koizumi 1972; Kwa & Liew 1975). Rickard and Outteridge (1974) showed that rabbits infected with *T. pisiformis* gradually developed delayed-hypersensitivity (DH) skin reactions to oncospheral antigen preparations known to stimulate protective immunity. However, when DH reactions were used as markers for identifying antigen during purification procedures, those antigens giving DH reactions did not stimulate protective immunity in rabbits (Rickard & Katiyar 1976). Spleen cells from normal or immune mice adhere to *Hymenolepis nana* oncospheres *in vitro*, but do not appear to cause any damage (Furukawa 1974). This is another area in which research is required, and *in vitro* cultivation methods may be useful here also. If methods for purification of T cell populations are refined, reconstitution experiments using the T-cell-deficient nude mouse should be valuable.

Experiments utilizing serial passage of *E. granulosus, E. multilocularis* and *T. crassiceps* metacestodes have shed some light on mechanisms which may destroy mature larval stages. Larvae generally become resistant to antibody attack early in their development (Campbell 1938; Heath 1973b; Musoke & Williams 1975; Mitchell, Goding & Rickard 1977) and experiments attempting to demonstrate a role for antibody in immunity to metacestode stages have given far less striking results than those described for oncospheres (Ali-Khan 1974; Musoke & Williams 1976; Siebert, Good & Simmons 1978a, 1978b). The latter authors described rapid changes in the surface of *T. crassiceps* larvae soon after they were implanted into immune hosts and they attributed this to antibody-dependent complement-mediated lysis. Similar results were obtained when larvae were incubated in immune serum *in vitro* (Siebert & Good 1979). There were qualitative differences in this "early" killing in different inbred strains of mice (Siebert & Good 1980). Normal serum induces lytic changes at the surface of other taeniid metacestodes (Chen 1950; Herd 1976; Rickard, Mackinlay, Kane, Matossian & Smyth 1977; Williams, Picone & Engelkirk 1980) and this has been attributed to activation of complement via the alternate pathway (Herd 1976; Rickard et al. 1977) so that results obtained with immune serum must be interpreted with caution.

Cellular defence mechanisms have been implicated in resistance to metacestode stages. Probably the most convincing demonstration of this was that antibody and macrophages collaborate in killing *E. multilocularis* protoscolices *in vitro* (Rau & Tanner 1976; Baron & Tanner 1977) and the authors suggested that these cells play a key role in preventing dissemination of the parasites *in vivo*. Effective encapsulation of parasites may also be important. Experiments with *M. corti* in mice suggest that encapsulation is T-cell-dependent (Pollaco et al. 1978) and Baron and Tanner (1976) showed that T-cell-suppressed mice failed to limit the metastasis and growth of *E. multilocularis*. Siebert and Good (1980) found that the encapsulation response of BALB/c and BDF1 mice to *T. crassiceps* larvae was very much more rapid than in C3H mice, with nearly 100% of the larvae being encapsulated within 3 weeks of infection. The capsule may serve to protect the parasite against some immune defences although antibody can apparently diffuse through it (Varela-Díaz & Coltorti 1972, 1973; Willms & Arcos 1977; Kwa & Liew 1978). Recent studies with *E. multilocularis* in mice suggest that in this parasite the laminated membrane provides a major barrier to immunoglobulin (Ali-Khan & Siboo 1981) and in *E. granulosus* this structure is known to be produced by the parasite and not by the host (Heath & Osborn 1976). Mast cells may play a part in host capsule formation by neutralizing parasite products toxic to the host cells (Siebert et al. 1979).

Rat peritoneal cells, especially eosinophils and mast cells, adhered rapidly to *T. taeniaeformis* strobilocerci incubated *in vitro* in normal or immune rat serum (Engelkirk, Williams & Signs 1981). The cells, especially eosinophils, caused severe damage to the distal tegument by engulfing tegumentary cytoplasm. Mast cells, but not eosinophils, underwent degranulation. The significance of these *in vitro* events to those which occur *in vivo* are not at all clear, because mature strobilocerci survive transplantation into the peritoneal cavity of rats (Musoke & Williams 1976).

Some experiments suggest that T-cell-mediated cytotoxicity may have a role to play in immunity to metacestodes. There is increased blastogenesis of both normal and immune lymphoid cells on exposure to antigen *in vitro* (Rickard & Outteridge 1974; Araj, Matossian & Malakian 1977; Dixon et al. 1978; Dixon et al. 1982). Thymus-cell-depleted mice are unable to control the dissemination of *E. multilocularis* and nude mice have been shown to be highly susceptible to infection with *T. taeniaeformis* by Mitchell, Goding and Rickard (1977) and *M. corti* by Mitchell, Marchalonis, Smith, Nicholas and Warner (1977). Anderson and Griffin (1979a, 1979b) found that resistance to *T. crassiceps* in young AS-strain rats was associated with the responsiveness of host lymphoid cells to polyclonal T cell activators, and that immunity was transferred adoptively with lymph node cells, but not with serum. Once again, the problem in interpreting these experiments is to distinguish between true cellular effects and T-cell-dependent antibody-mediated effects.

Clearly, our understanding of the relationship between the immune system of the host and established metacestodes is fragmentary. Research in this area is essential if methods for immunotherapy are ever to be developed. Also many of the pathological consequences of infection are due to destruction of mature larvae by the host defence system. The similarity between persistence of cestode larvae and the persistence and dissemination of tumours is striking and useful information on the host-tumour relationship may be gained by studying these parasites.

Evasion of the Immune Response

The survival of cestode larvae in the intermediate host is essential for transmission but the animal must not become so heavily infected that it dies before falling prey to the carnivorous definitive host. To ensure this, animals infected with larval taeniid cestodes display classical "concomitant" immunity, i.e. survival of mature larval stages in the face of almost absolute resistance to challenge infection. Schistosomes also display concomitant immunity (Capron, Aurialt, Mazingue, Capron & Torpier 1980; McLaren & Terry 1982; Sher & Benno 1982) and this group of parasites has been studied extensively. Mechanisms that have been suggested to play a part in protecting schistosomes from the host defence mechanisms include: masking by host antigens (Smithers & Terry 1969; Goldring, Clegg, Smithers & Terry 1976; McLaren & Terry 1982); molecular mimicry (Damian 1964, 1979); alterations in antigenicity (Samuelson, Sher & Caulfield 1980; Sher & Benno 1982); intrinsic changes in the resistance of the tegument (Kemp, Brown, Merritt & Miller 1980; Moser, Wassom & Sher 1980); enzymatic cleavage of host Ig (Aurialt, Quaiss, Torpier, Eisen & Capron 1981); complement activation (Machado, Gazzinelli, Pellegrino & Dias da Silva 1975). Few of these phenomena have been as intensively studied in larval taeniid cestodes, but information is available on some of them. A detailed consideration of this topic can be found in Rickard and Williams (1982).

Masking by host antigens

Host proteins have been demonstrated on the surface of larval cestodes including *E. granulosus* by Varela-Díaz and Coltorti (1973); *E. multilocularis* by Ali-Khan and Siboo (1981); *T. taeniaeformis* by Heath and Elsdon Dew (1972), Kwa and Liew (1978); *T. saginata* by Soulé, Remond and Chevrier (1979); *T. crassiceps* by Chernin (1977); and *T. solium* by Willms and Arcos (1977). These proteins may have been passively absorbed onto the parasite surface, or alternatively they may be antibodies bound to specific antigenic determinants. The concept of blocking antibody as a defence for larval cestodes against cell-mediated immune mechanisms of the host was advanced by Rickard (1974) and has had some support (Kwa & Liew 1978). Ali-Khan and Siboo (1981) found that all classes of antibody except IgA bound to the surface of *E. multilocularis* cysts but suggested (Ali-Khan & Siboo 1982) that IgM may compete more effectively than other Ig antibodies and thus sterically interfere with the attachment of Fc-receptor bearing cells. These authors further suggested that bound IgG_{2b} may block the Fc of IgG_{2a} from reacting with macrophages (Ali-Khan & Siboo 1982). All classes of antibody also bind to the surface of *T. crassiceps* larvae with IgM and IgG_1 predominating (Siebert, Blitz, Morita & Good 1981). Carefully controlled studies using specific antibody classes from immune sera in a combination of *in vitro* cultivation and *in vivo* implantation may further clarify the role of blocking antibody. Blocking antibody alone does not always provide total protection because many mature *T. crassiceps* larvae are destroyed when inoculated into immune hosts despite having host proteins on their surface (Siebert et al. 1978a, 1978b).

There is no evidence for the attachment of host red cell glycolipids or histocompatibility antigens to cestode larvae as has been shown to occur with schistosomes (Goldring et al. 1976; Sher, Hall & Vadis 1978).

Molecular mimicry

Initially proposed by Damian (1964), the concept of "molecular mimicry" or disguise by synthesizing host-like antigens on the parasite surface, has been supported to some extent (Capron, Biguet, Vernes & Afchain 1968). In recent studies, Willms, Merchant, Arcos, Sealy, Diaz and de Leon (1980) purified IgG from the surface of *T. solium* cysticerci and found that it showed no specificity for cysticercus antigens. This IgG may have been passively absorbed onto the cysticercus surface or else synthesized by the parasite itself. In a further experiment they found that one of the protein products produced following *in vitro* translation of parasite-derived RNA was precipitated by anti-porcine IgG. Unless the parasite RNA was contaminated with RNA from adherent host cells, this experiment constitutes striking evidence for molecular mimicry.

Alterations in antigenicity

It has already been stated that post-oncospheral stages rapidly outgrow their susceptibility to immune attack by the host (Campbell 1938; Heath 1973b; Musoke & Williams 1975; Mitchell, Goding & Rickard 1977). During this period there are profound alterations in the surface structure of the parasite (Furukawa et al. 1981; Rickard & Williams 1982; Engelkirk & Williams 1982). While these changes may reflect some fundamental physiological changes which increase resistance to attack by the host, they may also represent qualitative differences in antigenicity. There is evidence that *T. ovis* and *T. hydatigena* oncospheres possess some stage-specific antigenic determinants (Craig & Rickard 1981, 1982) although many components are shared with later developmental stages. The results of those experiments were not, however, conclusive and other explanations for the data are possible, e.g. oncospheral antigens may be internalized or in lesser proportion in the more advanced larval stages, thus giving apparent qualitative differences. In the mouse-*T. taeniaeformis* system, strobilocercus antigens immunize against infection with eggs and this apparently mitigates against stage specificity. However, it is not known which stage of the parasite is affected by the immune response to strobilocercus antigens, i.e. the oncosphere or later developmental stages. Studies in our laboratory using murine hybridoma antibodies raised against *T. taeniaeformis* oncospheres have produced clones of cells producing antibody with greater reactivity to oncospheral antigen than to strobilocercus or whole worm antigen, but in no case has antibody been totally oncosphere-specific (Lightowlers, Mitchell & Rickard, unpublished results).

Defence by the tegument

The tegument of cestodes is biologically very active (Lumsden 1975) and it is possible that antigenic modulation of the surface, such as patching and capping as described for *S. mansoni* (Kemp et al. 1980) could take place. In a recent publication Siebert et al. (1981) reported that they were unsuccessful in labelling living *T. crassiceps* larvae with fluorescent antibody because of the apparent ability of the parasites to shed aggregated immunoglobulins from their surface. Further investigations of this subject are necessary.

Modulation of host defences by substances released from parasites

The process of "fabulation" described by Eisen and Tallen (1977) and Aurialt et al. (1981), i.e. proteolytic cleavage of bound antibody to free the Fc portion and thus render the antibody ineffective in facilitating cell binding or complement fixation, has not been described for taeniid cestodes. *T. pisiformis* larvae release low molecular weight inhibitors of host proteolytic enzymes (Németh & Juhász 1980). Other parasite products have been shown to influence host cell mobilization and differentiation (Cook, Williams & Lichtenberger 1981), affect cell mobility (Goetzl & Austen 1977), reduce responsiveness to mitogens (Annen, Kohler, Eckert & Speck 1980), be cytotoxic (Mills, Coley & Williams 1984), and activate complement via the alternate pathway (Leid & Williams 1979). Each of these mechanisms in itself may be ineffective, but taken together they constitute a substantial attack on the defence system of the host.

Complement activation is the only one of these factors that has been studied in any detail. Anti-complementary substances are present in the tissue and cyst fluid of all metacestodes thus far examined and Williams et al. (1980) proposed a substantial role for decomplementation in the vicinity of larval cysts as a defence mechanism against killing by complement fixing antibody.

Immunosuppression

Evidence for immunosuppression during infections with larval taeniids is fragmentary. Factors in the serum of rabbits heavily infected with *T. pisiformis* inhibit lectin-mediated blast transformation of lymphocytes (Rickard, M.D. & Outteridge, P.M., unpublished results), as does a substance from *E. granulosus* cyst fluid (Annen et al. 1980). Polyclonal B cell activation has been suggested as a mechanism for immunosuppression in *T. solium* infection (Sulivan-Lopez, Sealy, Ramos, Melendro, Willms & Ortiz-Ortiz 1980). Recently Allan, Jenkins, Connor and Dixon (1981) showed that lymph node cells transferred from BALB/c mice infected with *E. granulosus* to syngeneic normal mice suppressed the response of recipients to sheep erythrocytes; the cell transfer innocula were shown to have depleted Thy-1 cells. Another example of immunosuppression is the impaired ability of mice infected with *T. crassiceps* to make antibodies to sheep red blood cells (Good & Miller 1976). The role of general immunosuppressive phenomena in larval survival needs clarification.

Immunopathology

Healthy larval cestodes living in their intermediate hosts are often of little clinical consequence other than due to mechanical interference because of their size. However, substantial pathological consequences can result from host-parasite interaction.

Allergic phenomena due to the interaction of antigen with reaginic antibodies are common in hydatid disease of humans. These can range from anaphylactic shock due to rupture of cysts, or sometimes without cyst rupture (Deenchin, Milenkov, Tevziev & Tashkova 1976), or repeated allergic episodes similar to angioneurotic oedema (Werczberger, Golhman, Wertheim, Gunders & Chowers 1979). Many other animal species also develop reaginic antibodies during larval cestode infections, e.g. pigs (Marler 1978), cattle (Moreira, Dos Santos & Oliveira 1978), sheep (Williams 1975; Schantz 1973), rabbits (Leid & Williams 1975) and rats (Leid & Williams 1974), but the significance of these in affecting the host is not known. They may even play a role in protective immunity (Leid & Williams 1979).

There is evidence that circulating immune-complexes can become deposited in renal glomeruli causing membranous immune-complex nephropathy in chronic echinococcosis in humans (Vialtel, Chenais, Desgeorges, Couderc, Micouin & Cordonnier 1981).

Neurocysticercosis in humans is the classic example of where an intense inflammatory reaction resulting from death of cysticerci causes the major clinical symptoms (Showramma & Reddy 1963). However, death and subsequent degeneration of cestode larvae is a common feature of several larval cestode infections in animals, e.g. *T. ovis* in sheep and *T. saginata* in cattle, and detailed studies of these may give important information on the pathogenesis of clinical cysticercosis in humans. In experimental hosts it has been possible to mimic death of the parasite, with its consequent inflammatory reaction, by damaging metacestodes with anthelmintic drugs (Verheyen, Vanparijs, Borgers & Thienpont 1978).

Immunization

Studies on immunization against egg-induced infections with larval taeniid cestodes have been extensively reviewed, both early work (Gemmell & Soulsby 1968; Gemmell & Macnamara 1972; Gemmell 1976; Gemmell & Johnstone 1977) and more recent findings (Clegg & Smith 1978; Williams 1979; Rickard & Williams 1982). It is not the intention to present a detailed review here, but to outline briefly the progress that has been made and the challenges that are still to be overcome.

All intermediate hosts of larval taeniid cestodes thus far examined have been able to be immunized against experimental infection with eggs. In general, oncospheral antigens have given the best results in domesticated animals, but antigens from a variety of larval stages are effective in laboratory animals. Experiments using antigens prepared during *in vitro* culture of living oncospheres have shown that:

1. Animals can be protected against experimental challenge infection using antigen in oil adjuvant (sheep, cattle, mice).
2. Similar injection protocols protect animals against infection acquired by grazing contaminated pasture (sheep, cattle).
3. Calves and lambs can be protected against infection early in life via colostral antibody from mothers immunized before parturition, and they can be immunized while under cover of colostral immunity.
4. Immunity in lambs to *T. ovis* persists for at least 12 months following a single injection of antigen in oil adjuvant.
5. An immunization schedule can be applied to animals reared under normal farm management conditions which will give significant levels of protection (cattle).
6. Cross-protection with antigen from heterologous species of cestodes is not as effective as immunization with homologous antigens.

Although this is an impressive list of achievements in vaccination studies, there are many problems to overcome before the development of commercial vaccines is possible. A major question that must be answered is the level of protection needed to achieve control in any given situation. Further studies along the lines of computer modelling described by Harris, Revfeim and Heath (1980) may provide information on this point.

The question of supply and cost are crucial in developing an economic vaccine. Recent work has shown that *in vitro* culture of oncospheres is not necessary to obtain protective antigens and that frozen, thawed and sonicated oncospheres are equally effective (Rajasekariah, Mitchell & Rickard 1980; Rajasekariah, Rickard & Mitchell 1980; Rickard & Brumley 1981). Even though this eliminates the problem of costly culture methods, it still does not solve the difficulty of supply. *E. granulosus* and *T. solium* eggs are dangerous to personnel required to handle them, and *T. solium* and *T. saginata* eggs are difficult to obtain in large quantities from humans. It is unlikely that dogs could be maintained on a large enough scale to supply the numbers of *T. ovis, T. hydatigena* or *E. granulosus* eggs needed for commercial vaccine production. The answers to these problems must lie in developing methods for large scale *in vitro* production of antigen.

Work in our laboratory has commenced with this end in view. Firstly it has been shown that the apparently insoluble membrane antigens can be solubilized with sodium deoxycholate (DOC) without losing their immunogenicity (Rajasekariah, Rickard, Mitchell & Anders 1982). Several approaches are being taken toward isolation of the "functional" antigen(s). A major problem is the minute amounts of material available to work with — 20,000 oncospheres yields only some 10 μg of protein. Murine hybridoma antibodies raised to *T. taeniaeformis* oncospheral antigens are being tested in passive immunization experiments. Several hybridoma antibodies are presently available (both IgM and IgG) which react strongly with oncospheral antigen in an *in vitro* ELISA test. If protective antibody is found it can be used to affinity purify functional antigen or as an immunological probe in recombinant DNA studies. Immunoprecipitation of radio-labelled antigen using sera from mice known to contain protective antibody is another avenue being explored to enable identification of "functional" antigens. Antigens are disassociated from the precipitating antibodies, separated in sodium dodecylsulphate polyacrylamide gel electrophoresis and visualized by autoradiography. Antigens obtained in this way can be tested for their ability to immunize mice.

Other methods which may be useful for further study are the *in vitro* culture of parasite cell lines (Sakamoto 1978) if the appropriate antigen containing cells can be identified, or the fusion of parasite cells with other tissue culture cell lines; Howell (1981) produced hybrid cells between *Fasciola hepatica* and a rat fibroblast cell line, and the progeny contained parasite protein.

Conclusion

This brief review of immunity and immune-related phenomena in infections with larval cestodes highlights the many advances that have been made in our understanding of this subject in recent years. However, it is quite apparent that more questions remain unanswered than answered in most areas. Modern biotechnology should allow systematic studies to be undertaken to address many of these problems. The potential of these organisms as models for studying basic immunological phenomena of the host-parasite relationship of tissue parasites is immense. In addition, the prospects for development of practical vaccines to assist in control programmes is clear. Hopefully, research in the next decade will yield answers to many of the questions posed here.

Acknowledgments

My appreciation to Barbara Chambers and Denise Heffernan for their expert typing and Rhonda Honey for reviewing the manuscript. Work carried out in this laboratory was supported by the Australian Meat Research Committee and the National Health and Medical Research Council of Australia.

References

Ali-Khan, Z. 1974. Host-parasite relationship in *Echinococcosis*. I. Parasite biomass and antibody response in three strains of inbred mice against graded doses of *Echinococcus multilocularis*. *Journal of Parasitology* **60**:231–35.

Ali-Khan, Z. and Siboo, R. 1981. *Echinococcus multilocularis*: Distribution and persistence of specific host immunoglobulins on cyst membranes. *Experimental Parasitology* **51**:159–68.

Ali-Khan, Z. and Siboo, R. 1982. *Echinococcus multilocularis*: Immunoglobulin and antibody response in C57BL/6J mice. *Experimental Parasitology* **53**:97–104.

Allan, D., Jenkins, P., Connor, R.J. and Dixon, J.B. 1981. A study of immunoregulation of BALB/c mice by *Echinococcus granulosus equinus* during prolonged infection. *Parasite Immunology* **3**:137–42.

Ambu, S. and Kwa, B.H. 1980. Susceptibility of rats to *Taenia taeniaeformis* infection. *Journal of Helminthology* **54**:43–44.

Anderson, M.J.D. and Griffin, J.F.T. 1979a. *Taenia crassiceps* in the rat. I.Differences in susceptibility to infection and development of immunocompetence in relation to age and host strain. *International Journal for Parasitology* **9**:229–33.

Anderson, M.J.D. and Griffin, J.F.T. 1979b. *Taenia crassiceps* in the rat. II. Transfer of immunity and immunocompetence with lymph node cells. *International Journal for Parasitology* **9**:235–39.

Annen, J., Kohler, P., Eckert, J. and Speck, S. 1980. Cytotoxic properties of *Echinococcus granulosus* cyst fluid. In *The host-invader interplay*, ed. H. Van den Bossche, 339–42. Amsterdam: Elsevier/North-Holland Biomedical Press.

Ansari, A. and Williams, J.F. 1976. The eosinophilic response of the rat to infection with *Taenia taeniaeformis*. *Journal of Parasitology* **62**:728–36.

Ansari, A., Williams, J.F. and Musoke, A.J. 1976. Antibody mediated secondary eosinophilic response to *Taenia taeniaeformis* in the rat. *Journal of Parasitology* **62**:737–40.

Araj, G.F., Matossian, R.M. and Malakian, A.H. 1977. The host response in secondary hydatidosis of mice. II. Cell mediated immunity. *Zeitschrift für Parasitenkunde* **52**:31–38.

Auriault, C., Quaiss, M.A., Torpier, G., Eisen, H. and Capron, A. 1981. Proteolytic cleavage of IgG bound to the Fc receptor of *Schistosoma mansoni* schistosomula. *Parasite Immunology* **3**:33–44.

Baron, R.W. and Tanner, C.E. 1976. The effect of immunosuppression on secondary *Echinococcus multilocularis* infections in mice. *International Journal for Parasitology* **6**:31–42.

Baron, R.W. and Tanner, C.E. 1977. *Echinococcus multilocularis* in the mouse; the *in vitro* protoscolicidal activity of peritoneal macrophages. *International Journal for Parasitology* **7**:489–95.

Brandt, J.R.A. and Sewell, M.M.H. 1981. Varying infectivity of *Taenia taeniaeformis* for rats and mice. *Veterinary Research Communications* **5**:187–91.

Butterworth, A.E., Vadis, M.A. and David, J.R. 1980. Mechanisms of eosinophil-mediated helminthotoxicity. In *The eosinophil in health and disease*, ed. A.A.F. Mahmoud and K.F. Austen, 253–73. New York: Grune and Stratton.

Campbell, D.H. 1938. The specific protective property of serum from rats infected with *Cysticercus crassicollis*. *Journal of Immunology* **35**:195–204.

Capron, A., Aurialt, C., Mazingue, C., Capron, M. and Torpier, G. 1980. Schistosome mechanisms of evasion. In *The host invader interplay*, ed. H. Van den Bossche, 217–25. Amsterdam:Elsevier/North-Holland Biomedical Press.

Capron, A., Biguet, J., Vernes, A. and Afchain, D. 1968. Structure antigènique des helminthes. Aspects immunologiques des relations hôte-parasite. *Pathologia Biologica Paris* **16**:121–38.

Capron, M., Capron, A., Torpier, G., Bazin, H., Bout, D. and Joseph, M. 1978. Eosinophil-dependent cytotoxicity in rat schistosomiasis. Involvement of IgG_{2a} antibody and the role of mast cells. *European Journal of Immunology* **8**:127–33.

Capron, A., Dessaint, J.P., Capron, M. and Bazin, H. 1975. Specific IgE antibodies in immune adherence of normal macrophages to *Schistosoma mansoni* schistosomules. *Nature (London)* **253**:474–75.

Chen, H.T. 1950. The *in vitro* action of rat immune serum on the larvae of *Taenia taeniaeformis*. *Journal of Infectious Diseases* **86**:205–13.

Chernin, J. 1977. Common host antigens in laboratory rats infected with the metacestodes of *Taenia crassiceps*. *Journal of Helminthology* **51**:215–20.

Chernin, J. 1981. The growth and antigenicity of the metacestodes of *Taenia crassiceps* in several different strains of rats. *Journal of Helminthology* **55**:209–22.

Clegg, J.A. and Smith, M.A. 1978. Prospects for the development of dead vaccines against helminths. *Advances in Parasitology* **16**:165–218.

Coleman, E.J. and de Salva, J.J. 1963. Mast cell responses to cestode infection. *Proceedings of the Society for Experimental Biology and Medicine* **112**:432–34.

Coman, B.J. and Rickard, M.D. 1975. The location of *Taenia pisiformis, Taenia ovis* and *Taenia hydatigena* in the gut of the dog and its effect on net environmental contamination with ova. *Zeitschrift für Parasitenkunde* **47**:237–48.

Cook, R.W., Williams, J.F. and Lichtenberger, L.M. 1981. Hyperplastic gastropathy in the rat due to *Taenia taeniaeformis* infection: Parabiotic transfer and hypergastrinemia. *Gastroenterology* **80**:728–34.

Craig, P.S. and Rickard, M.D. 1981. Anti-oncospheral antibodies in the serum of lambs experimentally infected with either *Taenia ovis* or *Taenia hydatigena*. *Zeitschrift für Parasitenkunde* **64**:169–77.

Craig, P.S. and Rickard, M.D. 1982. Antibody responses of experimentally infected lambs to antigens collected during *in vitro* maintenance of the adult, metacestode or oncosphere stages of *Taenia hydatigena* and *Taenia ovis* with further observations on anti-oncospheral antibodies. *Zeitschrift für Parasitenkunde* **67**:197–209.

Curtis, M.R., Dunning, W.F. and Bullock, F.D. 1933. Genetic factors in relation to the etiology of malignant tumours. *American Journal of Cancer* **17**:894–923.

Damian, R.T. 1964. Molecular mimicry: Antigen sharing by parasite and host and its consequences. *American Naturalist* **98**:129–49.

Damian, R.T. 1979. Molecular mimicry in biological adaptation. In *Host parasite interfaces*, ed. B.B. Nickel, 103–26. New York: Academic Press.

Deenchin, P., Milenkov, K.H., Tevziev, G. and Tashkova, M. 1976. Hydatidosis in autopsy data. *S'Vremennaya Meditsina* **27**:28–32.

Dixon, J.B.; Jenkins, P. and Allan, D. 1982. Immune recognition of *Echinococcus granulosus*. 1.Parasite-activated, primary transformation by normal murine lymph node cells. *Parasite Immunology* **4**:33–45.

Dixon, J.B., Jenkins, P., Allan, D. and Connor, R.J. 1978. Blastic stimulation of unprimed mouse lymphocytes by living protoscolices of *Echinococcus granulosus*: A possible connection with transplant immunity. *Journal of Parasitology* **64**:949–50.

Dow, C. and Jarrett, W.F.H. 1960. Age, strain and sex differences in susceptibility to *Cysticercus fasciolaris* in the mouse. *Experimental Parasitology* **10**:72–74.

Eisen, H. and Tallan, I. 1977. *Tetrahymena pyriformis* recovers from antibody immobilisation producing univalent antibody fragments. *Nature (London)* **270**:514.

Engelkirk, P.G. and Williams, J.F. 1982. *Taenia taeniaeformis* in the rat: Ultrastructure of the host-parasite interface on days 1–7 post infection. *Journal of Parasitology* **68**:620–33.

Engelkirk, P.G., Williams, J.F. and Signs, M.M. 1981. Interactions between *Taenia taeniaeformis* and host cells *in vitro*. I. Rapid adherence of peritoneal cells to strobilocerci. *International Journal for Parasitology* **11**:463–74.

Flisser, A., Pérez-Montfort, R. and Larralde, C. 1979. The immunology of human and animal cysticercosis: A review. *Bulletin of the World Health Organisation* **57**:839–56.

Froyd, G. 1960. Cysticercosis and hydatid disease of cattle in Kenya. *Journal of Parasitology* **46**:491–96.

Furukawa, T. 1974. Adherence reactions with mouse lymphoid cells against the oncosphere larvae of *Hymenolepis nana*. *Japanese Journal of Parasitology* **23**:236–49.

Furukawa, T., Niwa, A. and Miyazato, T. 1981. Ultrastructural aspects of immune damage to *Hymenolepis nana* oncospheres in mice. *International Journal for Parasitology* **11**:287–300.

Gemmell, M.A. 1976. Immunological responses and regulation of the cestode zoonoses. In *Immunology of human parasitic infections*, ed. S. Cohen and E. Sadun, 333–58. Oxford: Blackwell Scientific Publications.

Gemmell, M.A. and Johnstone, P.D. 1977. Experimental epidemiology of hydatidosis and cysticercosis. *Advances in Parasitology* **15**:311–69.

Gemmell, M.A. and Macnamara, F.N. 1972. Immune responses to tissue parasites II. Cestodes. In *Immunity to animal parasites*, ed. E.J.L. Soulsby, 235–72. London, New York: Academic Press.

Gemmell, M.A. and Soulsby, E.J.L. 1968. The development of acquired immunity to tapeworms and progress towards active immunization, with special reference to *Echinococcus* spp. *Bulletin of the World Health Organization* **39**:45–55.

Ginsberg, A., Cammeron, J., Goddard, W.B. and Grieve, J.M. 1956. Bovine cysticercosis with particular reference to East Africa. *East Africa Medical Journal* **33**:495–505.

Goetzl, E.J. and Austen, K.F.A. 1977. Cellular characteristics of the eosinophil compatible with a dual role in host defence in parasitic infections. *American Journal of Tropical Medicine and Hygiene* **26**:142–50.

Goldring, O.L., Clegg, J.A., Smithers, S.R. and Terry, R.J. 1976. Acquistion of human blood group antigens by *Schistosoma mansoni*. *Clinical and Experimental Immunology* **26**:181–87.

Good, A.H. and Miller, K.L. 1976. Depression of the immune response to sheep erythrocytes in mice infected with *Taenia crassiceps* larvae. *Infection and Immunity* **14**:449–56.

Harris, R.E., Revfeim, K.J.A. and Heath, D.D. 1980. Simulating strategies for control of *Echinococcus granulosus, Taenia hydatigena* and *T. ovis*. *Journal of Hygiene, Cambridge* **84**:389–404.

Heath, D.D. 1971. The migration of oncospheres of *Taenia pisiformis, T. serialis* and *Echinococcus granulosus* within the intermediate host. *International Journal for Parasitology* **1**:145–52.

Heath, D.D. 1973a. An improved technique for the *in vitro* culture of taeniid larvae. *International Journal for Parasitology* **3**:481–84.

Heath, D.D. 1973b. Resistance to *Taenia pisiformis* larvae in rabbits. II. Temporal relationships and the development phase affected. *International Journal for Parasitology* **3**:491–98.

Heath, D.D. and Elsdon-Dew, R. 1972. The *in vitro* culture of *Taenia saginata* and *Taenia taeniaeformis* larvae from the oncosphere, with observations on the role of serum for *in vitro* culture of larval cestodes. *International Journal for Parasitology* **2**:119–30.

Heath, D.D. and Lawrence, S.B. 1976. *Echinococcus granulosus*: Development *in vitro* from oncosphere to immature hydatid cyst. *Parasitology* **73**:417–23.

Heath, D.D. and Lawrence, S.B. 1981. *Echinococcus granulosus* cysts: Early development *in vitro* in the presence of serum from infected sheep. *International Journal for Parasitology* **11**:261–66.

Heath, D.D. and Osborn, P.J. 1976. Formation of *Echinococcus granulosus* laminated membrane in a defined medium. *International Journal for Parasitology* **6**:467–71.

Heath, D.D. and Pavloff, P. 1975. The fate of *Taenia taeniaeformis* oncospheres in normal and passively protected rats. *International Journal for Parasitology* **5**:83–88.

Herd, R.P. 1976. The cestocidal effect of complement in normal and immune sera *in vitro. Parasitology* **72**:325–34.

Howell, M.J. 1981. An approach to the production of helminth antigens *in vitro*: The formation of hybrid cells between *Fasciola hepatica* and a rat fibroblast cell line. *International Journal for Parasitology* **11**:235–42.

Kemp, W.M., Brown, P.R., Merritt, S.C. and Miller, R.E. 1980. Tegument associated antigen modulation by adult male *Schistosoma mansoni. Journal of Immunology* **124**:806–11.

Kwa, B.H. and Liew, F.Y. 1975. The role of cell-mediated immunity in *Taenia taeniaeformis* infections. *South East Asian Journal of Tropical Medicine and Public Health* **6**:483–87.

Kwa, B.H. and Liew, F.Y. 1978. Studies on the mechanism of long-term survival of *Taenia taeniaeformis* in rats. *Journal of Helminthology* **52**:1–6.

Lawrence, S.B., Heath, D.D., Parameter, S.N. and Osborn, P.J. 1980. Development of early larval stages of *Taenia ovis in vitro* using a cell monolayer. *Parasitology* **81**:35–40.

Leid, R.W. and Williams, J.F. 1974. The immunological response of the rat to infection with *Taenia taeniaeformis*. II. Characterization of reaginic antibody and an allergen associated with the larval stage. *Immunology* **27**:209–26.

Leid, R.W. and Williams, J.F. 1975. Reaginic antibody response in rabbits infected with *Taenia pisiformis. International Journal for Parasitology* **5**:203–8.

Leid, R.W. and Williams, J.F. 1979. Helminth parasites and the host inflammatory system. *Chemical Zoology* **11**:229–71.

Lloyd, S. and Soulsby, E.J.L. 1978. The role of IgA immunoglobulins in the passive transfer of protection to *Taenia taeniaeformis* in the mouse. *Immunology* **34**:939–45.

Lubinsky, G. 1964. Growth of the vegetatively propagated strain of larval *Echinococcus multilocularis* in some strains of Jackson mice and in their hybrids. *Canadian Journal of Zoology* **42**:1099–1103.

Lumsden, R.D. 1975. Surface ultrastructure and cytochemistry of parasitic helminths. *Experimental Parasitology* **37**:267–339.

Machado, A.J., Gazzinelli, G., Pellegrino, J. and Dias da Silva, W. 1975. *Schistosoma mansoni*: The role of complement C3-activating system in the cercaricidal action of normal serum. *Experimental Parasitology* **38**:20–29.

McLaren, D.J. and Terry, R.J. 1982. The protective role of acquired host antigens during schistosome maturation. *Parasite Immunology* **4**:129–48.

Marler, J.M. 1978. Antigeno intradermico para diagnostico de cysticercose. *Revista Saude Publica* **12**:23–25.

Mills, G.L., Coley, S.C. and Williams, J.F. 1984. Lipid and protein composition of the surface tegument from larvae of *Taenia taeniaeformis*. *Journal of Parasitology* **70**:197–207.

Mitchell, G.F., Goding, J.W. and Rickard, M.D. 1977. Studies on the immune response to larval cestodes in mice. I. Increased susceptibility of certain mouse strains and hypothymic mice to *Taenia taeniaeformis* and analysis of passive transfer of resistance with serum. *Australian Journal of Experimental Biology and Medical Science* **55**:165–86.

Mitchell, G.F., Marchalonis, J.J., Smith, P.M., Nicholas, W.L. and Warner, N.L. 1977. Studies on immune responses to larval cestodes in mice: Immunoglobulins associated with the larvae of *Mesocestoides corti*. *Australian Journal of Experimental Biology and Medical Science* **55**:187–211.

Mitchell, G.F., Rajasekariah, G.R. and Rickard, M.D. 1980. A mechanism to account for mouse strain variation in resistance to the larval cestode, *Taenia taeniaeformis*. *Immunology* **39**:481–89.

Moreira, W.S., Santos, A.F. Dos and Oliveira, Q.C. 1978. A intradermo-reacão de Casoni no diagnostico da hidatidose de bovinos. *Revista do Centro De Ciencias Rurais* **8**:97–103.

Moser, G., Wassom, D. and Sher, A. 1980. Studies of the antibody-dependent killing of schistosomula of *Schistosoma mansoni* employing haptenic target antigens. I. Evidence that the loss in susceptibility to immune damage undergone by developing schistosomula involves a change unrelated to the masking of parasite antigens by host molecules. *Journal of Experimental Medicine* **152**:41–53.

Musoke, A.J. and Williams, J.F. 1975. The immunological response of the rat to infection with *Taenia taeniaeformis*. V. Sequence of appearance of protective immunoglobulins and the mechanism of action of $7S\gamma_{2a}$ antibodies. *Immunology* **29**:855–66.

Musoke, A.J. and Williams, J.F. 1976. Immunological response of the rat to infection with *Taenia taeniaeformis*: Protective antibody response to implanted parasites. *International Journal for Parasitology* **6**:265–69.

Musoke, A.J. Williams, J.F. and Leid, R.W. 1978. Immunological response of the rat to infection with *Taenia taeniaeformis*:VII. The role of immediate hypersensitivity in resistance to reinfection. *Immunology* **34**:565–70.

Musoke, A.J., Williams, J.F., Leid, R.W. and Williams, C.S.F. 1975. The immunological response of the rat to infection with *Taenia taeniaeformis*. IV. immunoglobulins involved in passive transfer of resistance from mother to offspring. *Immunology* **29**:845–53.

Németh, I. 1970. Immunological study of rabbit cysticercosis. II. Transfer of immunity to *Cysticercus pisiformis* (Bloch 1780) with parenterally administered immune serum or lymphoid cells. *Acta Veterinaria Academiae Scientarium Hungaricae* **20**:69–79.

Németh, I. and Juhász, S. 1980. A trypsin and chymotrypsin inhibitor from the metacestodes of *Taenia pisiformis*. *Parasitology* **80**:433–46.

Okamoto, K. 1968. Effect of neonatal thymectomy on acquired resistance to *Hymenolepis nana* in mice. *Japanese Journal of Parasitology* **17**:53–59.

Okamoto, K. and Koizumi, M. 1972. *Hymenolepis nana*: Effect of antithymocyte serum on acquired immunity in mice. *Experimental Parasitology* **32**:56–61.

Olivier, L. 1962. Natural resistance to *Taenia taeniaeformis*. I. Strain differences in susceptibility of rodents. *Journal of Parasitology* **48**:373–78.

Orihara, M. 1962. Studies on *Cysticercus fasciolaris*, especially on differences of susceptibility among uniform strains of the mouse. *Japanese Journal of Veterinary Research* **10**:37–56.

Pollaco, S., Nicholas, W.L., Mitchell, G.F. and Stewart, A.C. 1978. T-cell dependent collagenous encapsulating response in the mouse liver to *Mesocestodes corti*. *International Journal for Parasitology* **8**:457–67.

Rajasekariah, G.R., Mitchell, G.F. and Rickard, M.D. 1980. *Taenia taeniaeformis* in mice: Protective immunization with oncospheres and their products. *International Journal for Parasitology* **10**:155–60.

Rajasekariah, G.R., Rickard, M.D. and Mitchell, G.F. 1980. Immunization of mice against infection with *Taenia taeniaeformis* using various antigens prepared from eggs, oncospheres, developing larvae and strobilocerci. *International Journal for Parasitology* **10**:315–24.

Rajasekariah, G.R., Rickard, M.D., Mitchell, G.F. and Anders, R.F. 1982. Immunization of mice against *Taenia taeniaeformis* using solubilized oncospheral antigens. *International Journal for Parasitology* **12**:111–16.

Rau, M.E. and Tanner, C.E. 1975. BCG suppresses growth and metastasis of hydatid infections. *Nature (London)* **256**:318–19.

Rau, M.E. and Tanner, C.E. 1976. *Echinococcus multilocularis* in the cotton rat: The *in vitro* protoscolicidal activity of peritoneal cells. *International Journal for Parasitology* **6**:195–98.

Reuben, J.M., Tanner, C.E. and Rau, M.E. 1978. Immunoprophylaxis with BCG of experimental *Echinococcus multilocularis* infection. *Infection and Immunity* **21**:135–39.

Rickard, M.D. 1974. Hypothesis for the long term survival of *Taenia pisiformis* cysticerci in rabbits. *Zeitschrift für Parasitenkunde* **44**:203–9.

Rickard, M.D. and Brumley, J.L. 1981. Immunization of calves against *Taenia saginata* infection using antigens collected by *in vitro* incubation of *T. saginata* oncospheres or ultrasonic disintegration of *T. saginata* and *T. hydatigena* oncospheres. *Research in Veterinary Science* **30**:99–103.

Rickard, M.D., Brumley, J.L. and Anderson, G.A. 1982. A field trial to evaluate the use of antigens from *Taenia hydatigena* oncospheres to prevent infection with *T. saginata* in cattle grazing on sewage-irrigated pasture. *Research in Veterinary Science* **32**:189–93.

Rickard, M.D. and Katiyar, J.C. 1976. Partial purification of antigens collected during *in vitro* cultivation of the larval stages of *Taenia pisiformis*. *Parasitology* **72**:269–79.

Rickard, M.D., Mackinlay, L.M., Kane, G.J., Matossian, R.M. and Smyth, J.D.

1977. Studies on the mechanism of lysis of *Echinococcus granulosus* protoscoleces incubated in normal serum. *Journal of Helminthology* **51**:221–28.

Rickard, M.D. and Outteridge, P.M. 1974. Antibody and cell-mediated immunity in rabbits infected with the larval stages of *Taenia pisiformis. Zeitschrift für Parasitenkunde* **44**:187–201.

Rickard, M.D., Rajasekariah, G.R. and Mitchell, G.F. 1981. Immunisation of mice against *Taenia taeniaeformis* using antigens prepared from *T. pisiformis* and *T. hydatigena* eggs or oncospheres. *Zeitschrift für Parasitenkunde* **66**:49–56.

Rickard, M.D. and Williams, J.F. 1982. Hydatidosis/Cysticercosis: Immune mechanisms and immunization against infection. *Advances in Parasitology* **21**:229–96.

Sakamoto, T. 1978. Development of echinococcal tissue cultured *in vitro* and *in vitro*. *Memoirs of the Faculty of Agriculture, Kagomisha University* **14**:109–15.

Samuelson, J.C., Sher, A. and Caulfield, J.P. 1980. Newly transformed schistosomula spontaneously lose surface antigens and C3 acceptor sites during culture. *Journal of Immunology* **124**:2055–57.

Schantz, P.M. 1973. Homocytotropic antibody to *Echinococcus* antigen in sheep with homologous and heterologous larval cestode infection. *American Journal of Veterinary Research* **34**:1179–81.

Sher, A. and Benno, D. 1982. Decreasing immunogenicity of developing schistosome larvae. *Parasite Immunology* **4**:101–7.

Sher, A., Hall, B.F. and Vadas, M.A. 1978. Acquisition of murine major histocompatibility complex gene products by schistosomula of *Schistosoma mansoni. Journal of Experimental Medicine* **148**:46–57.

Showramma, A. and Reddy, D.B. 1963. Silent cysticercosis of the brain. An analysis of five cases with special reference to histopathology. *Indian Journal of Pathology and Bacteriology* **6**:142–47.

Siebert, A.E. Jr, Blitz, R.R., Morita, C.T. and Good, A.H. 1981. *Taenia crassiceps*: Serum and surface immunoglobulins in metacestode infections of mice. *Experimental Parasitology* **51**:418–30.

Siebert, A.E. Jr and Good, A.H. 1979. *Taenia crassiceps*: Effect of normal and immune serum on metacestodes *in vitro*. *Experimental Parasitology* **48**:164–74.

Siebert, A.E. Jr and Good, A.H. 1980. *Taenia crassiceps*: Immunity to metacestodes in BALB/c and BDF1 mice. *Experimental Parasitology* **50**:437–46.

Siebert, A.E. Jr, Good, A.H. and Simmons, J.E. 1978a. Kinetics of primary and secondary infections with *Taenia crassiceps* metacestodes (Zeder 1800) Rudolphi 1810 (Cestoda:Cyclophyllidea). *International Journal for Parasitology* **8**:39–43.

Siebert, A.E. Jr, Good, A.H. and Simmons, J.E. 1978b. Ultrastructural aspects of early immune damage to *Taenia crassiceps* metacestodes. *International Journal for Parasitology* **8**:45–53.

Siebert, A.E. Jr, Good, A.H. and Simmons, J.E. 1979. Ultrastructural aspects of the host cellular immune response to *Taenia crassiceps* metacestodes. *International Journal for Parasitology* **9**:323–31.

Silverman, P.H. 1955. A technique for studying the *in vitro* effect of serum on activated taeniid hexacanth embryos. *Nature (London)* **176**:598–99.

Singh, B.B. and Rao, B.V. 1967. On the development of *Cysticercus fasciolaris* in albino rat liver and its reaction on the host tissue. *Ceylon Veterinary Journal* **15**:121–29.

Smithers, S.R. and Terry, R.J. 1969. Immunology of schistosomiasis. *Advances in Parasitology* **7**:41–93.

Soulé, C., Remond, M. and Chevrier, L. 1979. *Cysticercus bovis*: antigènes issus du cysticerque en survie. *Recueil de Médecine Veterinaire* **155**:889–94.

Sulivan-Lopez, J., Sealey, M., Ramos, C., Melendro, E.J., Willms, K. and Ortiz-Ortiz, L. 1980. B lymphocyte stimulation by parasitic organisms. In *Molecules, cells and parasites in immunology*, ed. C. Larralde, K. Willms, L. Ortiz-Ortiz and M. Sela, 113–24. New York: Academic Press.

Thompson, R.C.A. 1976. Inhibitory effect of BCG on development of secondary hydatid cysts of *Echinococcus granulosus*. *Veterinary Record* **99**:273.

Thompson, R.C.A. 1978. Aspects of speciation in *Echinococcus granulosus*. *Veterinary Parasitology* **4**:121–25.

Thompson, R.C.A. 1979. Biology and speciation of *Echinococcus granulosus* with special reference to Australia. *Australian Veterinary Journal* **55**:93–98.

Thompson, R.C.A. and Kumaratilake, L.M. 1982. Intraspecific variation in *Echinococcus granulosus*: The Australian situation and perspectives for the future. *Transactions of the Royal Society of Tropical Medicine and Hygiene* **76**:13–16.

Thompson, R.C.A., Penhale, W.J., White, T.R. and Pass, D.A. 1982. BCG induced inhibition and destruction of *Taenia taeniaeformis* in mice. *Parasite Immunology* **4**:93–99.

Turner, H.M. and McKeever, S. 1976. The refractory responses of the White Swiss strain of *Mus musculus* to infection with *Taenia taeniaeformis*. *International Journal for Parasitology* **6**:483–87.

Varela-Díaz, V.M. and Coltorti, E.A. 1972. Further evidence of the passage of host immunoglobulins into hydatid cysts. *Journal of Parasitology* **58**:1015–16.

Varela-Díaz, V.M. and Coltorti, E.A. 1973. The presence of host immunoglobulins in hydatid cyst membranes. *Journal of Parasitology* **59**:484–88.

Varute, A.T. 1971. Mast cells in cyst-wall of hydatid cyst of *Taenia taeniaeformis* (Batsch). *Indian Journal of Experimental Biology* **9**:200–203.

Verheyen, A., Vanparijs, O., Borgers, M. and Thienpont, D. 1978. Scanning electron microscopic observations of *Cysticercus fasciolaris* (= *Taenia taeniaeformis*) after treatment of mice with mebendazole. *Journal of Parasitology* **64**:411–25.

Vialtel, P., Chenais, F., Desgeorges, P., Couderc, P., Micouin, C. and Cordonnier, D. 1981. Membranous nephropathy associated with hydatid disease. *New England Journal of Medicine* **304**:610–11.

Wakelin, D. 1976. Genetic control of susceptibility and resistance. *Advances in Parasitology* **16**:217–308.

Werczberger, A., Golhman, J., Wertheim, G., Gunders, A.E. and Chowers, I. 1979. Disseminated echinococcosis with repeated anaphylactic shock treated with mebendazole. *Chest* **76**:482–84.

Williams, B.M. 1975. Homocytotropic antibodies to *Coenurus cerebralis* antigen in sheep. *British Veterinary Journal* **131**:361–63.

Williams, J.F. 1979. Recent advances in the immunology of cestode infections. *Journal of Parasitology* **65**:337–49.

Williams, J.F., Picone, J. and Engelkirk, P. 1980. Evasion of immunity by cestodes. In *The host-invader interplay*, ed. H. Van den Bossche, 205–16. Amsterdam: Elsevier/North Holland Biomedical Press.

Williams, J.F., Shearer, A.M. and Ravitch, M.M. 1981. Differences in susceptibility of rat strains to experimental infection with *Taenia taeniaeformis*. *Journal of Parasitology* **67**:540–47.

Willms, K. and Arcos, L. 1977. *Taenia solium*: Host serum proteins on the cysticercus surface identified by an ultrastructural immunoenzyme technique. *Experimental Parasitology* **43**:396–406.

Willms, K., Merchant, M.T., Arcos, L., Sealey, M., Diaz, S. and de Leon, L.D. 1980. Immunopathology of cysticercosis. In *Molecules, cells and parasites in immunology*, ed. C. Larralde, K. Willms, L. Ortiz-Ortiz, and M. Sela, 145–62. New York: Academic Press.

Yamashita, J., Ohbayashi, M., Kitamura, Y., Suzuki, K. and Okugi, M. 1958. Studies on *Echinococcosis*. VIII. Experimental *Echinococcosis multilocularis* in various rodents, especially on the difference of susceptibility among uniform strains of the mouse. *Japanese Journal of Veterinary Research* **6**:135–56.

12 Transmission of *Theileria peramelis* Mackerras, 1959 by *Ixodes tasmani*

D.J. Weilgama

Introduction

Four species of *Theileria* have been recorded from Australia, namely *T. mutans* from cattle (Dodd, 1910), *T. tachyglossi* (Priestly, 1915) from the echidna (*Tachyglossus aculeatus*), *T. ornithorhynchi* (Mackerras, 1959) from the platypus (*Ornithorhynchus anatinus*) and *T. peramelis* (Mackerras, 1959) from the short-nosed (*Isoodon macrourus*) and the long-nosed (*Peramelis nasuta*) bandicoots. Mackerras also detected theilerial parasites in a rat kangaroo (*Potorous tridactylus*) and assigned them tentatively to the latter species.

Apart from the species in cattle, very little information is available on the *Theileria* spp. of other animals. The morphology of *T. peramelis* has been described by Mackerras (1959), but no information is available on the life history or its pathogenicity. Mackerras (1959) however, suggested that one of four tick species commonly found on bandicoots could be responsible for its transmission. This paper reports transmission studies on *T. peramelis* using *Ixodes tasmani* and *Haemaphysalis humerosa*, two common ticks on bandicoots.

Materials and Methods

Hosts

Short-nosed bandicoots, *I. macrourus*, which were either trapped in the wild or reared in the laboratory from the pouch-young stage, were used in these studies. The trapping of animals was done in the suburbs of Brisbane, using wire mesh traps baited with sweet potato.

Maintenance

The bandicoots were housed in wire mesh cages placed over water troughs and kept in a temperature-controlled room (22°–24°C) with a 12 h light/dark cycle, in an insect-free building. They were fed with commercial rat pellets and water was provided in drip bottles. Special precautions were taken for rearing "parasite-free" bandicoots for transmission studies. Thus, once a mother with pouch-young was trapped, the latter, if fully grown (Mackerras & Smith 1960), was weaned and raised in a fly-proof room. The very young were left with the mother in the fly-proof room after the mother was cleaned free of ectoparasites.

Ticks

Two species, *I. tasmani* and *H. humerosa*, were used. "Clean" colonies of both species were maintained by feeding the immatures and adults on "clean" hand-reared bandicoots. "Clean" *I. tasmani* were raised from the progeny of a female obtained from a "clean" possum *Trichosurus vulpecula*, whilst "clean" *H. humerosa* were from a female obtained from a "clean" *I. macrourus.*

Feeding and rearing of ticks

The feeding of ticks was performed on restrained bandicoots. This was done by placing the bandicoot in a stainless steel box, 30 × 20 × 12 cm, with a steel grid as a lid. The ticks were applied with a brush after the animal had quietened down. After infestation, the bandicoot was returned to its cage and all unattached ticks present in the box were destroyed. The fed and unfed ticks were all held in glass vials, 2.5 × 2.5 cm, and maintained at 25°C at 90% relative humidity.

Examination of blood smears

Both thick and thin blood smears were made from the tail vein and stained with Giemsa. Thin smears were stained in 1:25 Giemsa solution at pH 7.2 (phosphate buffer), whilst thick smears were stained according to the method of Mahoney and Saal (1961).

Results

I. tasmani

Adults

Clean *I. tasmani* nymphs were fed on a naturally-infected bandicoot and the engorged nymphs allowed to moult. Five males and 5 females (30 days old) from this batch were fed on a hand-reared clean bandicoot (836B). Only 3 females engorged. Thick and thin smears were examined daily and *T. peramelis* was detected on the 23rd day following tick application. The peripheral parasitaemia was always low. It reached a maximum of 0.76% by the 7th day and the animal died on the 10th day. The cause of death could not be ascertained. Examination of impression smears from organs did not reveal Koch's blue bodies.

Nymphs

Clean *I. tasmani* larvae were fed on a naturally-infected bandicoot and allowed to moult. The nymphs, when 18–28 days old, were fed on 2 clean young bandicoots. Eighty-three nymphs were fed on each animal. Thirty-one engorged on one of these bandicoots, whilst only 5 engorged on the other. *T. peramelis* appeared in the peripheral blood smears of the bandicoots on the 13th and 28th days respectively following application.

The parasitaemias were very low. The highest recorded was 0.17% in the first bandicoot and the organisms remained in the circulation for 30 days.

Trans-ovarian transmission

Ten female *I. tasmani* that had engorged on 2 naturally-infected bandicoots were allowed to oviposit. At intervals, 25–40 eggs from each female were squashed and examined after staining with Giemsa. Examinations were made on days 1, 3 and 5 following commencement of oviposition and once every 5 days thereafter, until hatching occurred. No *Theileria* was detected. Larvae from these egg masses failed to infect the 2 young bandicoots on which they were allowed to feed.

H. humerosa

Adults

Clean nymphs of *H. humerosa* were fed on a naturally-infected bandicoot and allowed to moult. Eight males and 8 females from this batch, when 8–16 days old, were fed on a clean young bandicoot. Only 3 females engorged. *Theileria* was not detected in this animal.

A second experiment was done on another pouch-young bandicoot using 6 male and 5 female *H. humerosa*, previously obtained as engorged nymphs from an infected bandicoot, but no infection developed.

Discussion

I. peramelis was the most frequently encountered haemoprotozoan parasite among the short-nosed bandicoots. Nearly 50% (38 of 78) of the bandicoots caught were infected in nature but none showed any clinical illness. There were, however, signs of anaemia such as anisocytosis, marked polychromasia and immature cells in circulation. High parasitaemias were not detected in natural or experimental infections. The infections were benign as has been reported in infections due to *T. mutans* by Neitz (1957) and *T. ovis* by Neitz (1972). No Koch's blue bodies were detected but Maltese Cross forms were seen at the initial stages of experimental infections. Mackerras (1959) reported similar findings but failed to find cross forms.

The present study revealed that *I. tasmani* could transmit *T. peramelis* to the short-nosed bandicoot trans-stadially but not trans-ovarially. Both nymphs and adults were capable of transmitting the infection. The incubation period varied from 13–28 days. In trials with nymphs, the period was longer when only a few nymphs completed engorgement. Similar variations in incubation period have been reported by Lewis (1950) in studies with *T. parva*, where transmissions were done using only 1 tick. It is possible that these variations were related to the number of infective organisms inoculated.

There was no trans-ovarial transmission. Hereditary transmission has not been proved conclusively in theilerial infections. Although Ray (1950) has reported trans-ovarial transmission among *Hyalomma* ticks, others such as Daubney and Said (1951) failed to obtain transmission of *T. annulata* with *H. excavatum*. The attempts made to transmit the infection with *H. humerosa* were not successful. It is unlikely that the failure was due to the small number of adult ticks used, since a similar number of *I. tasmani* transmitted the infection. Further, theilerial infections have been transmitted to cattle using just 1 or few ticks (Lewis 1950; Daubney & Said 1951).

The epidemiology of theilerioses indicates some degree of species specificity. Neitz (1957) has shown that the distribution of *T. parva* is known to be closely related to the distribution of *Rhipicephalus* spp. and that of *T. annulata* to *Hyalomma* spp. In the present study, bandicoots were found to be the most favoured host for *I. tasmani. I. holocyclus* was another species that was found in similar numbers on bandicoots, whilst *H. humerosa*, although known as the bandicoot tick (Seddon 1968), was comparatively uncommon in the present survey. Judging by the incidence and distribution of *I. tasmani* and the high rate of theilerial infection among bandicoots, it could be assumed that *I. tasmani* is the natural vector of *T. peramelis*. It is known with *T. parva* that the density of vectors and the presence of susceptible cattle determine the degree of prevalence (Neitz 1957).

The trapped bandicoots were found to retain the infection for periods up to 10 weeks. Mackerras (1959) reported seeing parasites for 13 weeks after infecting by blood inoculation. It is thus possible that bandicoots remain carriers for long periods and act as a source of re-infection for ticks.

Acknowledgments

The author is grateful to Dr H.M.D. Hoyte, Head of the Department of Parasitology, University of Queensland, for the facilities provided, and to Dr D.E. Moorhouse for the guidance and constant supervision of the work. Thanks are also due to Mr G. Wolf for his technical assistance and to Mr L.M. Siddell, for the care of the experimental animals. The financial assistance of the Australian Development Bureau is gratefully acknowledged. This paper is a part of the Ph.D. thesis submitted to the University of Queensland, Australia.

References

Daubney, R. and Said, M.S. 1951. Egyptian fever of cattle. The transmission of *Theileria annulata* (Dschunkowsky & Luhz, 1904) by *Hyalomma excavatum*, Koch 1844. *Parasitology* **41**:249–60.

Dodd, S. 1910. Diseases in stock. *Annual Report of the Department of Agriculture, Queensland for 1909–10* pp.19–22.

Lewis, E.N. 1950. Conditions affecting the East Coast Fever parasite in ticks and in cattle. *East African Agricultural Journal* **16**:65–77.

Mackerras, M.J. 1959. The haematozoa of Australian mammals. *Australian Journal of Zoology* **7**:105–35.

Mackerras, M.J. and Smith, R.H. 1960. Breeding the short-nosed marsupial bandicoot, *Isoodon macrourus* (Gould), in captivity. *Australian Journal of Zoology* **8**:371–82.

Mahoney, D.F. and Saal, J.R. 1961. Bovine babesiosis: Thick blood films for the detection of parasitaemia. *Australian Veterinary Journal* **37**:44–47.

Neitz, W.O. 1957. Theilerioses, gonderioses and cytauxzoonoses: A review. *Onderstepoort Journal of Veterinary Research* **27**:275–430.

Neitz, W.O. 1972. The experimental transmission of *Theileria ovis* by *Rhipicephalus evertsi mimeticus* and *R. bursa. Onderstepoort Journal of Veterinary Research* **39**:83–86.

Priestly, H. 1915. *Theileria tachyglossi* (n.sp.) a blood parasite of *Tachyglossus aculeatus. Annals of Tropical Medicine and Parasitology* **9**:233–38.

Ray, H.N. 1950. Hereditary transmission of *Theileria annulata* infection in the tick *Hyalomma aegyptium* (Neum). *Transactions of the Royal Society of Tropical Medicine and Hygiene* **44**:93-104.

Seddon, H.R. 1968. *Diseases of domestic animals in Australia.* Part 3. (Revised by H.E. Albiston 1967). Service Publications. Department of Health, Australia. Veterinary Hygiene, No. 7.

13 Aspects of the Biology, Seasonality and Host Associations of *Haemaphysalis bancrofti, H. humerosa, H. bremneri* and *Ixodes tasmani* (Acari:Ixodidae)

A.C.G. Heath

Introduction

The Australian continent is home to about 52 species of Ixodidae (Roberts 1970) and except for those of veterinary or medical importance, e.g. *Boophilus microplus, Haemaphysalis longicornis, Rhipicephalus sanguineus* and *Ixodes holocyclus*, the remainder of the fauna has not been studied in detail. The notable exception comprises investigations (Smyth 1973; Bull & Smyth 1973) of reptile ticks of the genera *Aponomma* and *Amblyomma*.

The work reported here involved ticks collected during other investigations in which small mammals were trapped to provide material for laboratory studies on other tick species (Heath 1979, 1981). Enough of some less well known species were obtained to warrant study of some aspects of their biology. The most frequently encountered was *Haemaphysalis bancrofti*. Although its host relationships and vector potential were dealt with by Roberts (1963, 1970) little else is known of its biology. Similarly, the host associations and distribution of *H. bremneri* and *Ixodes tasmani* have been investigated (Roberts 1970) but these species are still incompletely known, despite their common occurrence in Australia. The remaining species, *H. humerosa*, was studied in some detail by Smith (1941) but additional information is presented here.

Material and Methods

Ticks were collected from small mammals caught in cage traps set out at about monthly intervals in Brookfield, a Brisbane suburb. Other hosts were trapped at various localities in Queensland.

For laboratory studies, only those ticks taken from animals trapped at Brookfield were used. The hosts were placed in large wire mesh cages over trays of water. Engorged ticks dropped into the water, were collected, dried and placed under the experimental conditions within 2–3 h of detachment. In the case of *H. bremneri*, larvae and nymphs were also reared in the laboratory. Eggs laid by 2 female ticks taken from a *Trichosurus vulpecula* trapped in the wild were left to hatch at 25°C and 2 mm Hg saturation deficit (s.d.). About 150 larvae were placed on each of 2 adult laboratory mice held in a cage over a tray of water. The nymphs that emerged from the engorged larvae were then fed on an *Isoodon obesulus*, also held over water.

Haemaphysalis bancrofti engorged larvae were exposed to temperatures of 18°, 25°, 28° and 32°C at 2 mm Hg s.d. Engorged nymphs of *H. bancrofti* and engorged larvae and nymphs of *H. humerosa, H. bremneri* and *I. tasmani* were exposed only at 25°C and 2 mm Hg s.d. The maintenance of constant temperatures and s.d.s for the laboratory studies is described in detail in Heath (1979).

Ticks collected from animals trapped at other sites in Queensland were placed in 70% ethanol. Full details of all collections can be found in Heath (unpublished Ph.D. thesis, University of Queensland, 1974), but a summary of hosts trapped and tick species associated with them is presented in table 13.1. Additional host records and seasonal data for the species dealt with here were obtained from Roberts (1960, 1963), and from collections made available by Dr Valerie Davies, Queensland Museum; W. Dowd, James Cook University, Townsville; P. Ferris, Department of Primary Industries, Townsville; and A. Walford-Huggins, Kamerunga, Cairns.

Table 13.1 Mammals trapped and/or examined in Queensland from which the named tick species were collected

	Tick species			
Host species	*H. bancrofti*	*H. humerosa*	*H. bremneri*	*I. tasmani*
Aepyprymnus rufescens	+	–	–	–
Hydromys chrysogaster	–	–	–	+
Isoodon obesulus	+	+	–	+
Macropus agilis	+	–	–	–
M. rufogriseus	+	–	–	–
Melomys sp.	–	+	–	+
Perameles nasuta	+	+	–	+
Rattus rattus	+	–	–	+
Trichosurus vulpecula	+	–	+	+
Uromys caudimaculatus	–	+	–	+

+ Species collected
- Species not collected

Results

Seasonal distribution

(a) *Haemaphysalis bancrofti*

The seasonal distribution of *H. bancrofti* larvae and nymphs at Brookfield is shown in figure 13.1. Data for adult ticks from other

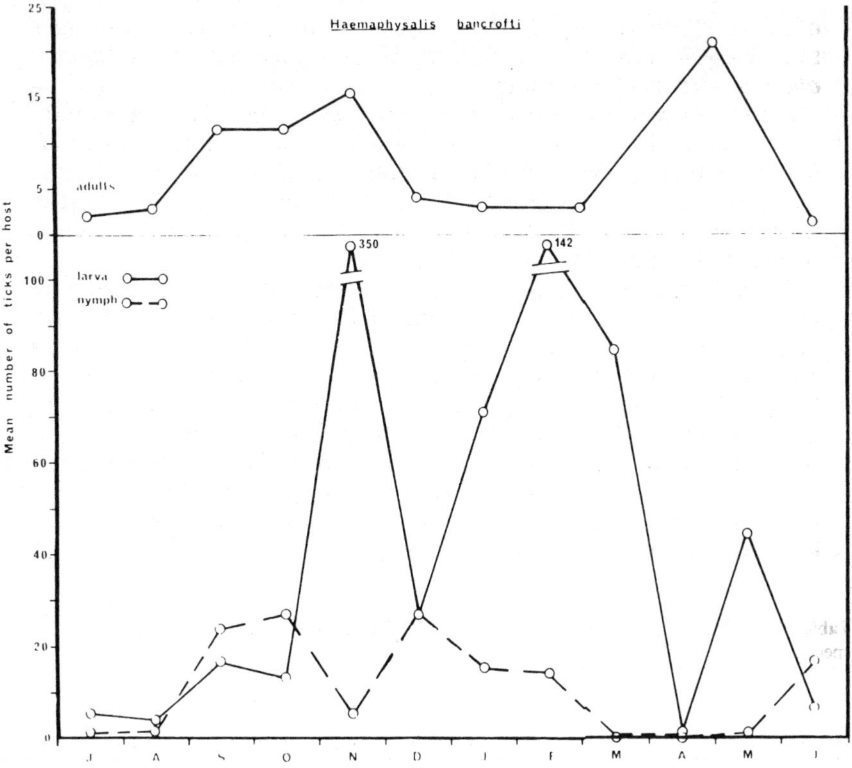

Figure 13.1 The seasonal activity pattern of the larvae and nymphs of *H. bancrofti* at Brookfield, together with data for adult ticks obtained from numerous sources.

sources are included. The numbers of each instar, the number of hosts examined and the number of infested hosts taken each month are shown in table 13.2.

(b) *Haemaphysalis humerosa, H. bremneri, I. tasmani*

The seasonal distribution of all instars of *H. humerosa, H. bremneri* and *I. tasmani* are shown in table 13.3. The data came from a wide variety of sources, with no guarantee that the total tick fauna from each host was obtained. For this reason, the mean intensity (Margolis, Esch, Holmes, Kuris & Schad 1982) of each tick species was not calculated, each instar merely being depicted as present or absent in a particular month.

Table 13.2 Instars of *H. bancrofti* taken at Brookfield each month between July 1973 and June 1974. Number of infested hosts in parentheses.

Instar	Month											
	J	A	S	O	N	D	J	F	M	A	M	J
Larva	16 (3)	12 (3)	83 (5)	13 (1)	350 (1)	27 (1)	71 (1)	284 (2)	170 (2)	1 (1)	89 (2)	19 (3)
Nymph	2 (2)	4 (3)	118 (5)	27 (1)	5 (1)	27 (1)	15 (1)	14 (1)	0	0	1 (1)	50 (3)
Adult	0	0	0	0	0	0	0	0	0	0	0	0
No. of hosts examined	3	5	5	1	1	2	1	3	2	4	3	3

Table 13.3 The seasonal distribution of all instars of *H. humerosa, H. bremneri* and *I. tasmani*

		Month											
Species	Instar	J	A	S	O	N	D	J	F	M	A	M	J
H. humerosa	larva	+	-	+	-	-	+	+	-	-	-	+	+
	nymph	+	+	+	-	-	+	+	-	-	-	+	-
	adult	+	+	+	+	-	-	+	-	-	+	+	+
H. bremneri	larva	+	+	-	-	-	-	-	-	-	-	-	-
	lymph	+	+	+	-	-	-	-	-	-	-	-	-
	adult	+	+	+	+	+	-	-	-	-	-	-	-
I. tasmani	larva	+	+	+	+	-	-	-	-	-	-	+	+
	nymph	+	+	+	+	+	+	+	-	-	-	+	+
	adult	+	+	+	+	+	+	+	+	+	+	+	+

\+ Present in that particular month.
\- Not present

Table 13.4 Mean (± S.E.) intensity data for each species and instar of tick together with the range. Number of observations in parentheses.

	Species			
Instar	*H. bancrofti*	*H. humerosa*	*H. bremneri*	*I. tasmani*
Larva	45.2 ± 86.6	21.1 ± 51.5	28.2 ± 15.3	5.4 ± 8.1
	1–350 (27)	1–175 (11)	1–36 (5)	1–32 (14)
Nymph	13.5 ± 12.1	45.9 ± 115.8	8.5 ± 5.5	3.7 ± 3.9
	1–43 (23)	1–354 (9)	1–14 (6)	1–16 (13)
Adult				
male	14 ± 0	38 ± 55.6	—	1 ± 0
	14 (1)	1–119 (4)	—	1 (2)
female	2 ± 0	14 ± 23.4	1.3 ± 0.6	4.6 ± 5.9
	2 (1)	1–49 (4)	1–2 (3)	1–15 (5)

Host associations

The mammal hosts from which all 4 species were collected are shown in table 13.1 The Wallabies *Macropus rufogriseus* and *M. parryi* were common at the Brookfield study area, and the Wallaroo *M. robustus* was seen occasionally, but none of these hosts in that area were examined for ticks.

Mean intensity data for each species and instar using both Brookfield collections and those from other Queensland localities are given in table 13.4. Material lent or given to me, or to be found in other publications was not included because there was no guarantee that such tick counts represented the total fauna on individual hosts.

Laboratory studies

(a) *Engorgement time, preoviposition and incubation periods*
The length of the attachment (engorgement) period for all instars, the preoviposition period of females and the incubation period for eggs are shown in table 13.5.

On the 2nd day after the mice had been infested with *H. bremneri* larvae, one mouse showed signs of ill-health and on the 5th day, when the last ticks had detached, both mice died.

(b) *Moulting times*
The time taken for the first moults to occur (premoult period) for *H. bancrofti* larvae at a range of temperatures and for *H. bancrofti* nymphs at 25°C are shown in table 13.6.

Premoult periods for larvae and nymphs of the other 3 species at 25°C are shown in table 13.7.

Table 13.5 The engorgement (attachment), preoviposition and incubation periods of 4 species of ticks, in days, determined from hosts trapped in the field. Number of observations in parentheses.

	Engorgement period§			Preoviposition period†	Incubation period†
Species	Larva	Nymph	Adult		
H. bancrofti	4–7 (8)	3–9 (5)	ND	ND	ND
H. humerosa	4–6 (3)	5–7 (2)	7 (2)	4 (2)	24–27 (2)
H. bremneri	4 (1)	3 (1)	10–17 (2)	3 (1)	No hatch
I. tasmani	3 (1)	ND	6 (1)	ND	No hatch

† Determined at 25°C and 2 mm Hg s.d.
§ Determined at room temperature
ND Not determined

Table 13.6 The premoult period (mean ± S.D.) in days for the larvae and nymphs of *H. bancrofti* at various temperatures

	Larvae				Nymphs
	18°C	25°C	28°C	32°C	25°C
Premoult period (mean ± S.D.)	32 ± 0	12.5 ± 1.25	11 ± 0	8 ± 0	14.6 ± 1.2
Range (days)	–	11–15	–	–	13–17
No. of ticks used	53	563	79	28	242
No. moulted	17	476	71	21	162
No. parasitized (*Hunterellus* sp.)	0	26	0	0	62
No. of experiments	1	21	1	1	22

Table 13.7 The premoult period (mean ± S.D.) in days for the larvae and nymphs of 3 species of tick at 25°C

	H. bremneri		*H. humerosa*		*I. tasmani*	
	larvae	nymphs	larvae	nymphs	larvae	nymphs
Premoult period (mean ± S.D.)	12.3 ± 1.2	15.9 ± 1.9	10 ± 0	13 ± 1.4	-	14.8 ± 1.6
Range (days)	11–14	13–19	-	12–14	-	12–16
No. of ticks used	499	68	100	37	55	10
No. moulted	438	51	97	28	0	8
No. parasitized (*Hunterellus* sp.)	0	1	0	0	0	2
No. of experiments	9	8	2	2	2	5

Discussion

An understanding of the seasonal pattern of activity of tick species is best achieved by regular sampling of the fauna in a small area and then comparing with results obtained in other areas. In this way differences in the seasonal patterns due to climate can be recognized.

In the present study, seasonal data for all species (although to a lesser extent for *H. bancrofti*) were drawn from records obtained from a diverse range of geographical and climatic regions. For this reason the results express the "average" seasonal activity of each species, a sum total of local patterns, rather than reflecting a local biotype.

Data collected at Brookfield and from Roberts (1960, 1963) show that each instar of *H. bancrofti* is present all year round. The continuing presence of all instars makes it likely that more than one generation is produced in a year. In fact, the data of Roberts (1960, 1963) show that there may be two peaks of adult activity (fig. 13.1) and the bimodality in larval activity at Brookfield may represent two age cohorts.

The absence of adult ticks from the Brookfield material may reflect host selection by this instar, in that collection of adults is invariably from Macropodidae (Heath 1974, thesis cited above) or large domestic animals (Roberts 1963). This illustrates a principle discussed by Macleod (1975) which suggests that, because of their behaviour or physiology, small mammals are unsuitable hosts for adult *H. bancrofti*, which may have sensory requirements differing from those of larval and nymphal instars.

The patterns of seasonal activity of the other two *Haemaphysalis* species extend over shorter periods than that of *H. bancrofti* and reflect either environmental requirements that differ from those of *H. bancrofti* or indicate that more extensive sampling is required. *Haemaphysalis bremneri* is almost exclusively taken from *Trichosurus vulpecula* (Roberts 1970; and present study) and closer examination of the ecology of this host may help explain the apparent restricted seasonal activity of the tick. *Trichosurus vulpecula* were taken sufficiently frequently at Brookfield to suggest that the seasonal activity pattern of *H. bremneri* as shown here is close to actuality.

The seasonal activity of *H. humerosa* as recorded by Smith (1941) is similar to that in the present study, except that Smith found neither larvae nor nymphs in December or February. He considered that his delay in getting hosts to the laboratory before larvae and nymphs detached may have explained their absence in those months. Smith (1941) suggested that two generations of *H. humerosa* were produced each year, and the presence of larvae in both spring and winter (table 13.3) may indicate that two age cohorts are present. *Haemaphysalis humerosa* is most commonly found on Peramelidae and rodents and is restricted to the northern half of Australia (Roberts 1970), which has a well-defined wet season. The extent to which such weather patterns affect host behaviour and *H. humerosa* life history has yet to be determined.

The year-round occurrence of *I. tasmani* adults is an indication that other instars are likely to be active in all seasons (as with *H. bancrofti*) despite their apparent absence in some months (table 13.3). The fact that *I. tasmani* is abundant, has a wide distribution and a large range of hosts (Roberts 1970) would support this view. Furthermore, the apparent absence can be explained by small samples in some months, or that small instars of ticks (particularly larvae) are often easily overlooked when a host is being searched, particularly in the field.

The response of *H. bancrofti* engorged larvae to a range of temperatures was similar to that seen by Heath (1981) in *H. longicornis* except that the premoult period for *H. bancrofti* at 18°C is about 8 days longer than for *H. longicornis*.

Moulting times for the other 3 species showed that the interspecific response of larvae and nymphs when held at 25°C were similar except that *H. humerosa* larvae and nymphs took a shorter time to moult at 25°C than comparable instars in other species. The data of Smith (1941) for *H. humerosa* moulting times were obtained at 14.8° and 21°C. Under these conditions, larvae moulted in 33 days at the lower temperature and nymphs in 39 days. Both stages moulted in 14 days at 21°C. The shorter moulting times in the present study are in accordance with the higher temperature used.

The s.d. used (2 mm Hg) in the laboratory experiments was sufficient to maintain a suitable water balance at 25°C in the larvae and nymphs of all species as mortality was generally low (3%–12.3% and 0%–24.3% respectively). At 18°C, mortality of *H. bancrofti* larvae exceeded 62%, comparable with *H. longicornis* larvae (Heath 1981) and was related to the longer time spent at 18°C before moulting, which lead to a water loss (Heath 1979, 1981).

The high mortality of *I. tasmani* larvae at 25°C is difficult to explain and may have been due to asphyxiation while the ticks were in water immediately following detachment. This may have also been a non-specific mortality factor for the other species, in contrast to the more easily attributable deaths caused by a parasitic wasp, *Hunterellus* sp. (tables 13.6 & 13.7) (Doube & Heath 1975). The studies reported here, although not comprehensive, provide a framework for further investigation and generate questions. For example, why should a tick species be active all year? Does this indicate that it finds a wide range of hosts suitable or is tolerant of a wide range of climatic and ecological conditions? *Haemaphysalis bancrofti* and *I. tasmani* have both a wide host and geographic range (Roberts 1970, table 13.1) and both occur throughout the year.

By contrast, do ticks that are active for only a small part of the year have a restricted host range or are they tolerant of only a narrow range of environmental conditions?*Haemaphysalis bremneri* satisfied these criteria, with *H. humerosa* intermediate between the two extremes.

Finally, do temperature and humidity preferences explain, reflect or support these apparent biological differences? It is here that further investigations are needed. Not all the species in this study were exposed to a range of temperatures and humidities and it is only by comparative studies of the type carried out by Heath (1979, 1981) that ecological requirements can be characterized and used to understand the complexities of tick life histories.

Acknowledgments

I am grateful for the assistance of those persons mentioned in the text who provided study material. Thanks are also due to Dr D.E. Moorhouse for advice and constructive criticism during this study which was carried out in the Department of Parasitology, University of Queensland. The work was funded by a New Zealand Government Study Award and by the Australian Meat Research Committee (Grant UQ–15).

References

Bull, M. and Smyth, M. 1973. The distribution of three species of reptile ticks, *Aponomma hydrosauri* (Denny), *Amblyomma albolimbatum* Neumann, and *Amb. limbatum* Neumann. II. Water balance of nymphs and adults in relation to distribution. *Australian Journal of Zoology* **21**:103–10.

Doube, B.M. and Heath, A.C.G. 1975. Observations on the biology and seasonal abundance of an encyrtid wasp, a parasite of ticks in Queensland. *Journal of Medical Entomology* **12**:443–47.

Heath, A.C.G. 1979. The temperature and humidity preferences of *Haemaphysalis longicornis, Ixodes holoyclus* and *Rhipicephalus sanguineus* (Ixodidae): Studies on eggs. *International Journal for Parasitology* **9**:33–39.

Heath, A.C.G. 1981. The temperature and humidity preferences of *Haemaphysalis longicornis, Ixodes holocyclus* and *Rhipicephalus sanguineus* (Ixodidae): Studies on engorged larvae. *International Journal for Parasitology* **11**:169–75.

Macleod, J. 1975. Apparent host selection by some African tick species. *Oecologia* **19**:359–70.

Margolis, L., Esch, G.W., Holmes, J.C., Kuris, A.M. and Schad, G.A. 1982. The use of ecological terms in parasitology (Report of an ad hoc committee of the American Society of Parasitologists). *Journal of Parasitology* **68**:131–33.

Roberts, F.H.S. 1960. A systematic study of the Australian species of the genus *Ixodes* (Acarina:Ixodidae). *Australian Journal of Zoology* **8**: 392–485.

Roberts, F.H.S. 1963. A systematic study of the Australian species of the genus *Haemaphysalis* Koch (Acarina:Ixodidae). *Australian Journal of Zoology* **11**:35–80.

Roberts, F.H.S. 1970. *Australian Ticks*. Melbourne: CSIRO.

Smith, D.J.W. 1941. Studies on the epidemiology of Q fever. 7. The biology of *Haemaphysalis humerosa* Warburton and Nuttall (Acarina:Ixodidae) in Queensland. *Australian Journal of Experimental Biology and Medical Science* **19**:73–75.

Smyth, M. 1973. The distribution of three species of reptile ticks, *Aponomma hydrosauri* (Denny), *Amblyomma albolimbatum* Neumann, and *Amb. limbatum* Neumann. I. Distribution and hosts. *Australian Journal of Zoology* **21**:91–101.

14 Species Segregation: Competition or Reinforcement of Reproductive Barriers?

K. Rohde and R.P. Hobbs

Introduction

The question of the ecological and evolutionary importance of competition has been central in population ecology and evolutionary theory. Many ecologists, until recently, considered interspecific competition to be the most important or even the only factor responsible for niche segregation.

During the last years, a renewed discussion on the factors affecting community structure has developed, and many ecologists now believe that interspecific competition is only one of several such factors (see review by Lewin 1983a, 1983b; with respect to parasites Rohde 1977, 1978, 1979, 1980a; Strong 1983). The importance of reinforcement of reproductive barriers has been discussed only rarely. Brown and Wilson (1956) pointed out that reinforcement (prevention of "gamete wastage" due to inferior hybrids produced by cross-fertilization of two closely related species) may be an important factor responsible for one aspect of ecological segregation, that is, character displacement. Miller (1967) in particular, drew attention to the fact that some examples of character displacement which had been explained by interspecific competition, may be explained as well or better by reinforcement. Rohde (1976, 1977, 1980a, 1980b, 1982) discussed the relative importance of competition and reinforcement for niche segregation in parasites and concluded that the latter may be more important (see also Sogandares-Bernal 1959; Martin 1969). A decision on the relative importance of the two interspecific factors based on the quantitative evaluation of data has not been made because most ecological systems are too complex for such an evaluation.

Ectoparasites of marine fish are an ideal and simple model to study ecological questions for several reasons, among which are the possibility of examining the body surface and gills of fish accurately and quantitatively in a short time, easy availability of fish in large numbers, and frequent occurrence on one individual of several and sometimes congeneric species of parasites (Rohde 1981). The resources used by these parasites, mainly food and space for attachment, can be analysed easily. Parasite species which use the same resources and live in the same microhabitat, for instance on the gills, but differ in their phylogenetic relationship, may differ in their degree of niche segregation and thus may give a clue as to whether competition or reinforcement is responsible for segregation (Rohde 1980b, 1982).

Materials and Methods

The following fish were examined: 122 *Scomber scombrus* at Helgoland, North Sea (*Kuhnia scombri* recorded in only 85 and cysts in 33 specimens); 37 *Trigla lucerna* at Helgoland; 120 *Gadus morhua* at Helgoland (*Clavella* recorded in only 45 and cysts in 5 specimens); 14 *Scomberomorus commerson* at Heron Island, Great Barrier Reef (gills only); 98 *Scomber japonicus* (body surface examined in only 50 specimens) and 18 *Oligoplites saliens*, both in São Paulo State, Brazil. Fish were usually killed immediately after capture and dropped into 10% formalin. Parasites were mapped on drawings of the fish and gills.

Gill parameters considered for the quantitative evaluation are: gill number, longitudinal section of the gill, external or internal filament, gill arch or filament, basal or distal parts of filaments (for details see Rohde 1978).

Several indices for niche overlap have been proposed (e.g. May 1975). The following indices are used in the following:

(1) percent similarity (Pielou 1969)

$$PS_{A,B} = 100 \sum_{i=1}^{n} \min(P_{iA}, P_{iB}),$$

where $PS_{A,B}$ = percent similarity of A and B, P_{iA} = proportion of species A in the ith microhabitat, P_{iB} = proportion of species B in the ith microhabitat, n = total number of microhabitats;

(2) a modified and asymmetrical percent similarity index (newly proposed)

$$O_{A,B} = \frac{100A}{N_A} \sum_{i=1}^{k} \min(Q_{iA}, Q_{iB}),$$

$$O_{B,A} = \frac{100B}{N_B} \sum_{i=1}^{k} \min(Q_{iA}, Q_{iB}),$$

where $O_{A,B}$ = overlap of A with B, A = number of individuals of species A in those k microhabitats in which B also occurs; N_A = total number of individuals of species A in all microhabitats; Q_{iA} and Q_{iB} = quotient of the number of individuals of species A and B respectively in microhabitat i and the total number of individuals of each species in the k microhabitats in which they co-occur.

The following comparisons are made:

(1) overlap between congeneric pairs with overlap between all non-congeners (CC vs CN,NN);

(2) overlap between congeneric pairs with overlap between non-congeners excluding those which do not co-occur with congeners (CC

vs CN). The non-parametric Wilcoxon's 2-sample test was performed on the data.

The figures of the copulatory organs are based on camera lucida drawings of whole mounts.

Results

The 6 fish species contained 35 species of ectoparasites, 19 of these occurring in congeneric species pairs or triplets on one host species. The genera include polyopisthocotylean Monogenea: *Kuhnia, Vallisia, Gotocotyla, Pseudothoracocotyla, Pricea, Microcotyle, Grubea*; monopisthocotylean Monogenea: *Udonella*; didymozoid Trematoda: *Nematobothrium*; Copepoda: *Caligus, Metacaligus, Pseudocycnoides, Clavellisa, Clavella, Neobrachiella, Lernaeocera, Lernentoma*; and unknown cysts. Species names are given in figures 14.1–14.6. Didymozoid trematodes and cysts, although endoparasites, are included because they live close to the surface and are therefore potential "competitors" of ectoparasites. *Caligus pelamydis* occurs in the mouth cavity and on the gills of *Scomber scombrus* and *S. japonicus; Caligus brevicaudatus* occurs on the surface, *C. diaphanus* in the mouth cavity and on the gills, *Neobrachiella impudica* in the mouth cavity, and *N. bispinosa* between

1								
18.851	2							
53.885	7.791	3						
0.000	0.000	0.000	4					
0.000	0.000	0.000	0.000	5				
35.007	46.654	18.396	0.000	0.000	6			
0.000	0.000	0.000	2.500	0.000	0.000	7		
40.713	29.450	40.885	0.000	0.000	32.417	0.000	8	
11.254	10.528	5.000	73.333	0.000	9.748	0.000	13.281	9

	1	2	3	4	5	6	7	8	9
1	100.000	18.851	42.385	0.000	0.000	28.881	0.000	40.713	47.493
2	18.851	100.000	3.583	0.000	0.000	45.633	0.000	29.450	35.183
3	63.904	44.949	100.000	0.000	0.000	37.873	0.000	52.991	55.556
4	0.000	0.000	0.000	100.000	0.000	0.000	2.321	0.000	87.143
5	0.000	0.000	0.000	0.000	100.000	0.000	0.000	0.000	0.000
6	38.761	49.780	10.766	0.000	0.000	100.000	0.000	40.351	46.538
7	0.000	0.000	0.000	46.429	0.000	0.000	100.000	0.000	0.000
8	40.713	29.450	34.983	0.000	0.000	30.263	0.000	100.000	66.016
9	6.332	4.691	2.778	73.333	0.000	5.429	0.000	8.802	100.000

Figure 14.1 Symmetrical (*top*) and asymmetrical (*bottom*) percent similarity indices. Indices for congeners underlined. *Scomber japonicus*. 1 — *Kuhnia* sp. I. 2 — *Kuhnia* sp. II. 3 — *Grubea* sp. 4 — *Nematobothrium* sp. I. 5 — *Nematobothrium* sp. II. 6 — *Nemateobothrium* sp. III. 7 — *Clavellisa scombri*. 8 — cysts. 9 — *Caligus pelamydis*.

1				
35.834	2			
0.000	0.000	3		
13.793	4.454	10.345	4	

	1	2	3	4
1	100.000	29.341	0.000	81.059
2	55.362	100.000	0.000	49.070
3	0.000	0.000	100.000	100.000
4	11.181	2.538	10.345	100.000

Figure 14.2 As fig. 14.1. *Scomber scombrus*. 1 — cysts. 2 — *Kuhnia scombri*. 3 — *Kuhnia* sp. 4 — *Caligus pelamydis*.

1					
7.432	2				
0.000	0.000	3			
0.000	0.785	0.000	4		
0.000	78.534	0.000	0.000	5	
0.000	0.000	100.000	0.000	0.000	6

	1	2	3	4	5	6
1	100.000	55.336	0.000	0.000	0.000	0.000
2	4.237	100.000	0.000	0.555	78.534	0.000
3	0.000	0.000	100.000	0.000	0.000	100.000
4	0.000	62.643	0.000	100.000	0.000	0.000
5	0.000	100.000	0.000	0.000	100.000	0.000
6	0.000	0.000	100.000	0.000	0.000	100.000

Figure 14.3 As fig. 14.1. *Trigla lucerna*. 1 — *Lernentoma asellina*. 2 — *Caligus diaphanus*. 3 — *Caligus brevicaudatus*. 4 — *Neobrachiella bispinosa*. 5 — *Neobrachiella impudica*. 6 — *Udonella caligorum*.

1				
14.332	2			
0.000	0.000	3		
19.538	66.797	0.000	4	
26.443	46.707	0.000	51.662	5

	1	2	3	4	5
1	100.000	55.160	0.000	56.051	37.027
2	8.634	100.000	0.000	70.412	42.968
3	0.000	0.000	100.000	0.000	0.000
4	12.331	61.963	0.000	100.000	47.481
5	20.570	52.199	0.000	55.533	100.000

Figure 14.4 As fig. 14.1. *Oligoplites saliens*. 1 — *Vallisia* sp. II. 2 — *Microcotyle* sp. 3 — *Vallisia* sp. I. 4 — *Caligus oligoplitsi*. 5 — *Metacaligus rufus*.

1			
4.237	2		
1.626	1.507	3	
0.000	0.000	0.000	4

	1	2	3	4
1	100.000	5.103	54.832	0.000
2	2.895	100.000	36.877	0.000
3	0.892	0.586	100.000	0.000
4	0.000	0.000	0.000	100.000

Figure 14.5 As fig. 14.1. *Gadus morhua*. 1 — cysts. 2 — *Lernaeocera branchialis*. 3 — *Clavella sciatherica*. 4 — *Clavella iadda*.

1						
63.592	2					
46.252	63.389	3				
17.947	39.011	50.187	4			
15.221	16.312	15.895	13.684	5		
25.144	36.170	38.393	28.846	0.000	6	
14.521	18.750	12.416	9.900	2.127	3.309	7

	1	2	3	4	5	6	7
1	100.000	63.592	46.252	16.360	8.014	12.753	28.708
2	63.592	100.000	63.389	37.255	13.022	24.938	44.315
3	46.252	63.389	100.000	42.765	10.009	33.104	23.222
4	25.204	42.986	63.060	100.000	5.843	18.636	17.652
5	52.472	79.833	62.379	42.698	100.000	0.000	15.492
6	48.888	68.947	86.225	49.697	0.000	100.000	43.795
7	9.865	12.179	8.770	5.296	1.601	1.884	100.000

Figure 14.6 As fig. 14.1. *Scomberomorus commerson*. 1 — *Gotocotyla bivaginalis*. 2 — *Pricea multae*. 3 — *Gotocotyla secunda*. 4 — *Pseudothoracocotyla indica*. 5 — *Pseudothoracocotyla gigantica*. 6 — *Pseudocycnoides armatus*. 7 — *Caligus* sp.

the pharyngeal plates and on the gills of *Trigla lucerna; Nematobothrium* sp. I and II occurs in the mouth cavity of *Scomber japonicus; Metacaligus rufus* occurs in the mouth cavity and on the gills, and *Caligus oligoplitsi* on the gills (and elsewhere ?) of *Oligoplites saliens; Clavella iadda* occurs on the fins, *C. sciatherica* on the body and gills of *Gadus morhua*. *Udonella caligorum* lives on *Caligus brevicaudatus*. All other species are found on the gills only.

Data for the symmetrical and asymmetrical indices are presented in figures 14.1–14.6. Wilcoxon's 2-sample test performed on the asymmetrical indices showed that congeners overlap less than non-congeners (df = ∞, CC vs CN: $t = 2.6550$, $P < 0.01$; CC vs CN,NN: $t = 2.8705$, $P < 0.01$). Some authors assume that *Clavella sciatherica* and

C. iadda on *Gadus morhua* belong to one species (Kabata 1979). Wilcoxon's test was therefore also carried out excluding all species on *Gadus morhua*. The results were: CC vs CN: $t=2.4090$, $P<0.02$, CC vs CN,NN: $t=2.6127$, $P<0.01$.

The results were not significant when percent similarity indices were used (CC vs CN: $t=1.6519$, ns; CC vs CN,NN: $t=1.7856$, ns; excluding *Gadus morhua*: CC vs CN: $t=1.4906$, CC vs CN,NN: $t=1.6067$, ns).

Discussion

Much information is lost in the symmetrical percent similarity index. Thus, cysts on the gills of *Scomber scombrus* have a much larger microhabitat than the monogenean *Kuhnia scombri*. Both species overlap, but whereas the whole microhabitat of the monogenean is also used by cysts, only a very small part of the microhabitat of cysts is also used by the monogenean. This difference in the relative overlap of both species is taken into consideration in the asymmetrical index, but not in the symmetrical index. The former therefore appears to be a more reliable indicator of species overlap and the conclusion is justified that congeners overlap significantly less than non-congeners.

The reason usually given for a clearer segregation of congeneric species is that related species use more similar resources than unrelated species and that competition is consequently more severe. However, all parasites discussed use the same resources, i.e. space for attachment and food provided by the host. All polyopisthocotylean Monogenea, whether related or not, feed on blood, as shown by many histochemical and EM studies (e.g. Rohde 1980c). The other species feed either on blood, tissue fluids, mucus or epithelial cells or a combination of these. All these food sources are in unlimited supply as long as the host is alive, i.e. they do not represent limiting factors. Furthermore, requirements for blood on the gills cannot be responsible for spatial niche segregation because it can be obtained from all parts of the gills. Epithelial cells etc. on the body surface or in the mouth cavity are also available from many parts. Hence, food requirements cannot be responsible for spatial segregation.

This leaves only one niche dimension for which the parasites may "compete" with resulting spatial segregation, that is space for attachment. "Competition" for space, however is not more severe in related species. For example, the trematode *Nematobothrium* sp. III on *Scomber japonicus* forms large cyst-like swellings on the gill filaments which make these filaments unsuitable for attachment of the monogenean *Kuhnia* or any other parasite, and occupation of a site by the copepod *Neobrachiella impudica* does not permit occupation by the copepod *Caligus* or any other parasite, whether related or not, on *Trigla lucerna*.

In summary, then, greater segregation of congeneric species cannot be explained by greater competition, it must be the consequence of reinforcement of reproductive barriers. This conclusion is supported by morphological observations.

The copulatory apparatus of all species of Monogenea and Didymozoa which are clearly segregated from congeneric species on the same host, have almost identical copulatory organs (examples figs. 14.7–14.9). Those

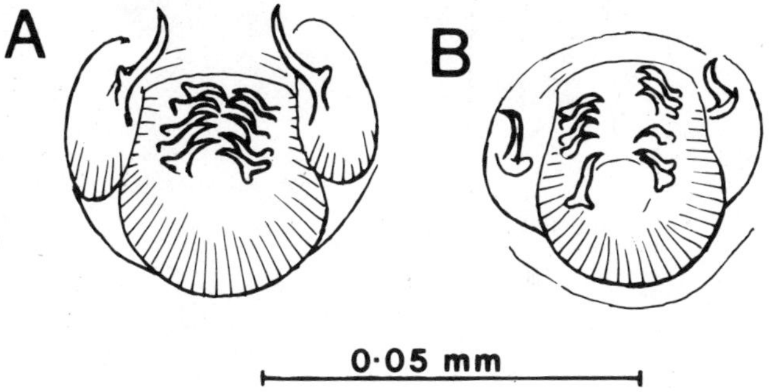

Figure 14.7 Male copulatory organ of *Kuhnia* sp. I (*A*) and *K.* sp. II (*B*) from *Scomber japonicus* in Brazil.

species with considerable overlap, i.e. *Gotocotyla* spp. and *Pseudothoracocotyla* spp., have very similar copulatory organs but some distinct differences. Thus, *Gotocotyla bivaginalis* has two sucker-like vaginae, and *G. secunda* has only one; *Pseudothoracocotyla indica* has a small cirrus papilla which is absent in *P. gigantica*. The latter species is also much larger (16 mm maximum length) than the former (5.5 mm) (Rohde 1976). If these species, i.e. all species occurring on *Scomberomorus commerson*, are excluded from the statistical evaluation, larger t- values are obtained (asymmetrical test, CC vs CN: $t = 2.9498$, $P < 0.01$; C vs CN, NN: $t = 3.1485$, $P < 0.01$; *Gadus morhua* also excluded [see above]: CC vs CN: $t = 2.7659$, $P < 0.01$; CC vs CN,NN: $t = 2.8795$, $P < 0.01$).

Data from the literature show that congeneric species of Monogenea sometimes have widely overlapping microhabitats. Thus, the monopisthocotylean monogeneans *Lamellodiscus acanthopagri, L. squamosus* and *L. major* co-occur on *Acanthopagrus australis* in the same gill microhabitats (Roubal 1981), and the monopisthocotyleans *Diplectanum aequans* and *D. laubieri* overlap widely on the gills of *Dicentrarchus labrax* (Lambert & Maillard 1974, 1975). These species have completely different copulatory organs (fig 14.10). It seems that these species employ a morphological instead of a spatial mechanism of reinforcing reproductive segregation.

A critical examination of data from the literature used to demonstrate interspecific competition leading to site segregation in parasites infecting the same host species, shows that well segregated species often are congeners. According to Schad (1963), 8 species of the nematode genus

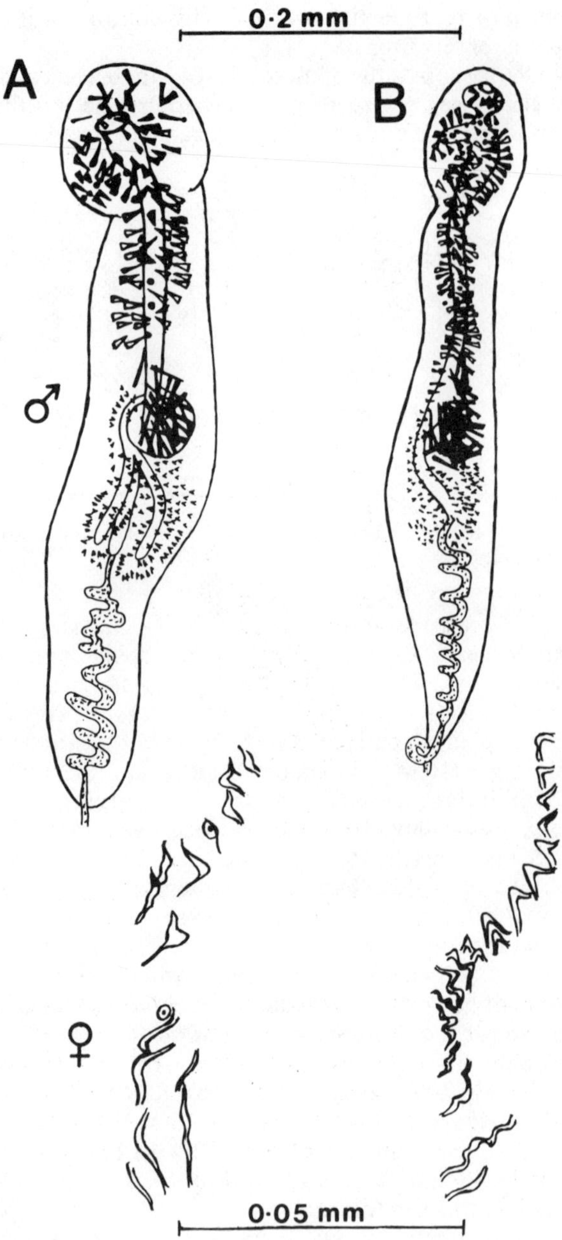

Figure 14.8 Male copulatory organ and vaginal sclerites of *Vallisia* sp. II (*A*) and *V*. sp. I (*B*) from *Oligoplites saliens* in Brazil.

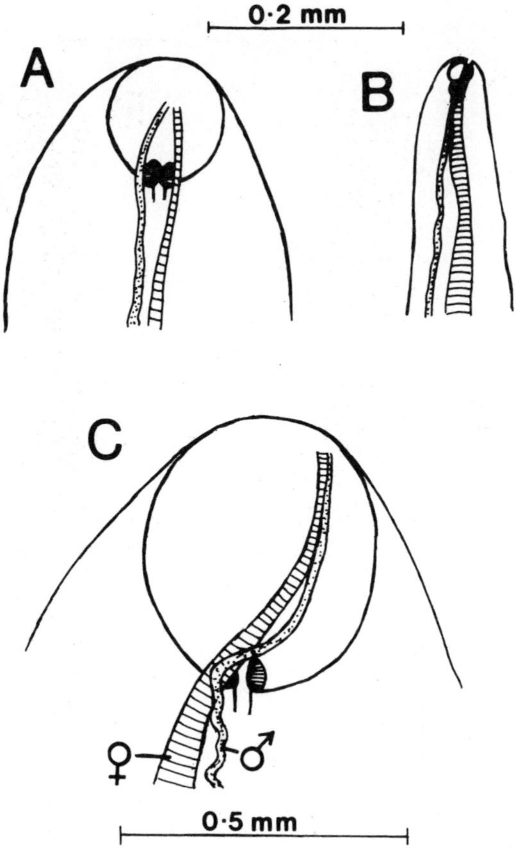

Figure 14.9 Anterior parts of body with terminal genital ducts and gonopores of *Nematobothrium* sp. I (*A*), *N.* sp. II (*C*) and *N.* sp. III (*B*) from *Scomber japonicus* in Brazil.

Tachygonetria in the colon of the turtle *Testudo graeca* were spatially segregated to varying degrees, and Uglem and Beck (1972) described spatial segregation in the acanthocephalans *Neoechinorhynchus cristatus* and *N. crassus*. In none of these cases was the possibility considered that reinforcement of reproductive barriers could be responsible. Only Sogandares-Bernal (1959) and Martin (1969) suggested, without giving evidence, that reinforcement was the decisive factor for segregation of trematodes studied by them.

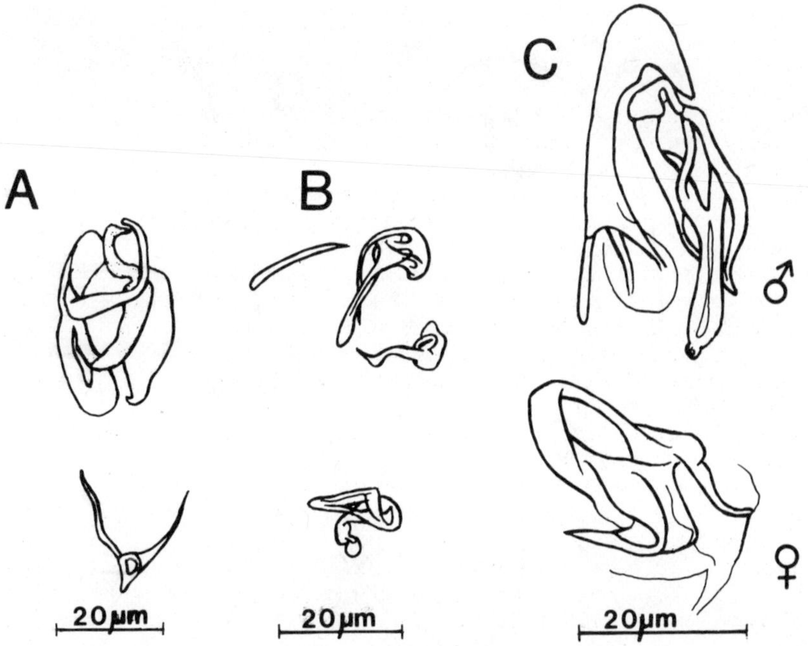

Figure 14.10 Male copulatory and vaginal sclerites of *Lamellodiscus ancanthopagri* (*A*), *L. squamosus* (*B*) and *L. major* (*C*) from *Acanthopagrus australis* in northern New South Wales. Modified from Roubal (1981).

References

Brown, W.L. Jr and Wilson, E.O. 1956. Character displacement. *Systematic Zoology* **5**:49–64.

Kabata, Z. 1979. *Parasitic Copepoda of British fishes*. London: Royal Society.

Lambert, A. and Maillard, C. 1974. Parasitisme branchial simultané par deux espèces de *Diplectanum* Diesing, 1858 (Monogenea, Monopisthocotylea) chez *Dicentrarchus labrax* (L., 1758) (téléostéen). *Comptes-Rendus hebdomadaires des séances, Academie des Sciences, Paris, ser. D* **279**:1345–47.

Lambert, A. and Maillard, C. 1975. Repartition branchiale de deux monogenes: *Diplectanum aequans* (Wagener 1857) Diesing, 1858 et *Diplectanum laubieri* Lambert et Maillard, 1974 (Monogenea, Monopisthocotylea) parasites simultanes de *Dicentrarchus labrax* (téléostéen). *Annales de Parasitologie humaine et comparée* **50**:691–99.

Lewin, R. 1983a. Santa Rosalia was a goal. *Science* **221**:636–38.

Lewin, R. 1983b. Predators and hurricanes change ecology. *Science* **221**:737–40.

Martin, D.R. 1969. Lecithodendriid trematodes from the bat *Peropteryx kappleri* in Colombia, including discussion of allometric growth and significance of ecological isolation. *Proceedings of the Helminthological Society of Washington* **36**:250–60.

May, R.M. 1975. Some notes on estimating the competition matrix, α. *Ecology* **56**:737–41.

Miller, R.S. 1967. Pattern and process in competition. *Advances in Ecological Research* **4**:1–74.

Pielou, E.C. 1969. *An introduction to mathematical ecology*. New York, London, Sydney, Toronto: Wiley-Interscience.

Rohde, K. 1976. Monogenean gill parasites of *Scomberomorus commersoni* Lacépède and other mackerel on the Australian east coast. *Zeitschrift für Parasitenkunde* **51**:49–69.

Rohde, K. 1977. A non-competitive mechanism responsible for restricting niches. *Zoologischer Anzeiger* **199**:164–72.

Rohde, K. 1978. Latitudinal gradients in species diversity and their causes. II. Marine parasitological evidence for a time hypothesis. *Biologisches Zentralblatt* **97**:405–18.

Rohde, K. 1979. A critical evaluation of intrinsic and extrinsic factors responsible for niche restriction in parasites. *American Naturalist* **114**:648–71.

Rohde, K. 1980a. Warum sind ökologische Nischen begrenzt? Zwischenartlicher Antagonismus oder innerartlicher Zusammenhalt? *Naturwissenschaftliche Rundschau* **33**:98–102.

Rohde, K. 1980b. Comparative studies on microhabitat utilization by ectoparasites of some marine fishes from the North Sea and Papua New Guinea. *Zoologischer Anzeiger* **204**:27–63.

Rohde, K. 1980c. Some aspects of the ultrastructure of *Gotocotyla secunda* and *Hexostoma euthynni. Angewandte Parasitologie* **21**:32–48.

Rohde, K. 1981. Niche width of parasites in species-rich and species-poor communities. *Experientia* **37**:359–61.

Rohde, K. 1982. *Ecology of marine parasites*. Brisbane: University of Queensland Press.

Roubal, F.R. 1981. The taxonomy and site specificity of the metazoan ectoparasites on the black bream, *Acanthopagrus australis* (Günther), in northern New South Wales. *Australian Journal of Zoology, Supplement* **84**:1–100.

Schad, G.A. 1963. Niche diversification in a parasitic species flock. *Nature* **198**:404–6.

Sogandares-Bernal, F. 1959. Digenetic trematodes of marine fishes from Gulf of Panama and Bimini, British West Indies. *Tulane Studies in Zoology* **7**:69–117.

Strong, D.R. Jr. 1983. Natural variability and the manifold mechanisms of ecological communities. *American Naturalist* **122**:636–60.

Uglem, G.L. and Beck, S.M. 1972. Habitat specificity and correlated aminopeptidase activity in the acanthocephalans *Neoechinorhynchus cristatus* and *N. crassus. Journal of Parasitology* **58**:911–20.

15 A Systems Approach to Ecological Research in Applied Parasitology

R.W. Sutherst

Introduction

Scientific research presents difficulties on different levels. Firstly, the analysis of isolated biological processes requires specific techniques which are provided by the traditional scientific education. Until recently less attention has been paid to a second difficulty, that of developing overall research strategies to define and analyse complex systems. How does an ecologist or epidemiologist, for example, go about analysing or solving an economically important parasite problem of domestic stock? Some systematic framework is needed and the approach must be based on ecological theory. In addition some rational means of setting priorities is required.

There are few comprehensive system studies of the ecology, epidemiology or management of livestock parasites. Suitable quantitative techniques have been developed in two areas. Entomologists were the first to use computer modelling to understand complex ecological systems (Holling 1964; Watt 1968). They also developed systematic ways of organizing ecological research programmes (Clark, Geier, Hughes & Morris 1967; Sutherst, Wharton & Utech 1978) and economic analyses of pest problems (Southwood & Norton 1973). Meanwhile, in the second area, veterinarians were starting to apply computers to animal disease problems (e.g. James & Ellis 1980).

During the 1960s and 1970s, the cattle tick was causing serious concern in Australia due to problems with chemical control (Wharton & Roulston 1970). An ecological study of the tick was started, primarily as a basis for understanding the role of resistant cattle in tick control (Sutherst & Wharton 1973; Wharton, Utech & Sutherst 1973). A systematic, large-scale programme continued for a decade (e.g. Sutherst & Dallwitz 1979; Sutherst, Norton, Barlow, Conway, Birley & Comins 1979; Maywald, Dallwitz & Sutherst 1980; Norton, Sutherst & Maywald 1983) using systems techniques to guide the programme and to analyse the data. That project has now ended and a similar one on the buffalo fly has just begun. This new study provides an opportunity to assess the generality and utility of the systems approach.

In this paper I outline a systems approach to the analysis of the ecology and control of the cattle tick, *Boophilus microplus*, in Australia and then explore its application to the buffalo fly, *Haematobia irritans exigua*, and other parasites.

The Systems Approach Applied to Cattle Tick

A framework for ecological analysis of parasite problems is shown in figure 15.1. Two basic relationships in pest control are (A) the damage function and (B) the control function. These relationships are integrated (C) for analysis of the pest parasite problem and finally the findings are implemented (D) in the industry.

A. Damage

The type and amount of damage caused by a parasite depends on the damage per parasite (d) and its relationship to population density per animal (N). If each individual causes the same loss at high or low densities the total damage D is simply dN. Estimation or prediction of N requires sampling or an understanding of the population dynamics of the parasite.

Damage per tick

Cattle ticks cause losses in live weight and reduced milk yields. Literature on the effect of *B. microplus* on the live weights of cattle was reviewed by Sutherst and Utech (1981). Recent studies by Johnston, Haydock and Leatch (1981), Sutherst, Maywald, Kerr and Stegeman (1983) and Mellor, O'Rourke and Waters (1983) have extended those observations. Irrespective of tick density, each female tick completing engorgement causes a loss (d) of about 0.6–2.0 g. Males are assumed to have a negligible effect as they feed to a much smaller extent. The effects on milk yield have not yet been quantified. Ticks also act as vectors for *Babesia* spp. and *Anaplasma* spp. The feeding lesions produce scar tissue that disfigures the surface grain of animal hides and so reduces their value. Severe tick infestations also cause hair loss and even open wounds which are attractive to the screw-worm fly in many other countries. In addition, there are often large opportunity costs associated with tick control which uses resources that could be used profitably on other activities.

Tick numbers and population dynamics

The numbers of ticks (N) per individual host are determined primarily by geographical location and the host's breed, sex and nutritional status. The distribution of ticks over the body of the host varies greatly with the tick species (e.g. Kaiser, Sutherst & Bourne 1982) as does the frequency distribution of tick numbers in a herd of animals. The breed of host is another major factor affecting the frequency distribution (Utech, Wharton & Kerr 1978) which is usually skewed, with most ticks being found on a small proportion of hosts.

In order to predict changes in tick numbers and the variation in damage to the host, the aims of the population ecologist are (fig. 15.2) to define the determinants of long-term average (equilibrium) population size, its

(A) Damage		(B) Control
A(I) Damage relationships	*A(II) Pest numbers and population dynamics*	*Identify control methods and define their effectiveness*
(i) type of damage	(i) Define pest distribution by geographical survey	Prevent development (e.g. habitat modification; predators; parasites)
For each type of damage, define (ii) quantity of damage per pest individual;	(ii) Define pest abundance and population dynamics by	
	(a) analysing population processes:	Prevent host finding (e.g. traps; pasture spelling; anti-tick pasture plants; pheromone baits)
any change in damage/pest individual with pest density; the extent of compensatory growth of the host; economic value of products	— freeliving development — host finding — parasitic feeding and mating	Prevent feeding or mating (e.g. host resistance; sterile males; chemicals)
	(b) longterm census in representative habitats	
	(c) Modelling the processes to simulate natural populations	Define management demands; costs (short and longterm); adverse consequences

(C) Management Model

Integrate: damage and control functions
population dynamics
economic relationships

Design robust integrated control strategy

(D) Implementation

Awareness campaign
Education campaign

Figure 15.1 Procedure for the analysis and control of parasite problems.

short-term fluctuations and the stability of the populations (i.e. their ability to return to equilibrium size after perturbation).

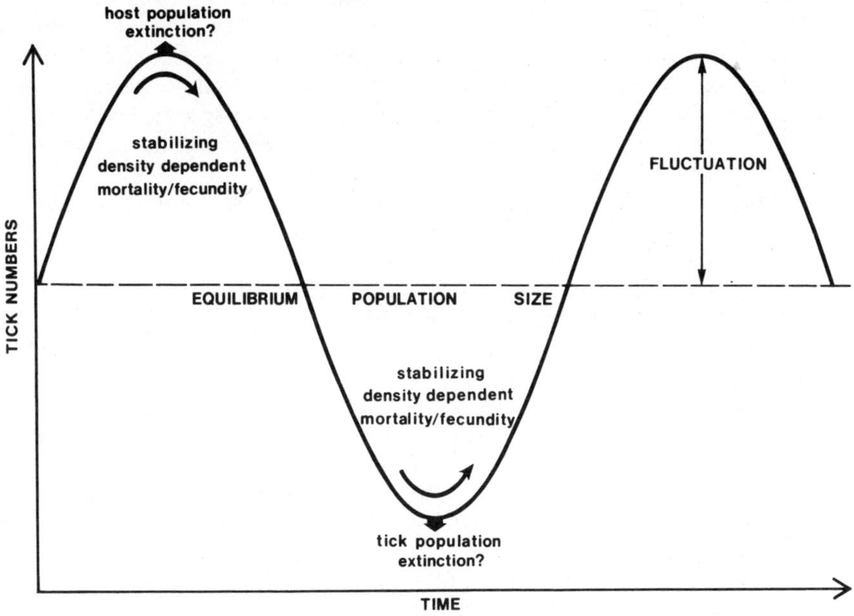

Figure 15.2 The characteristics of animal populations to be explained by ecologists.

The methods used to achieve these aims are to divide the life system into small components, to understand each and then to reassemble them into a functional model system (Holling 1964). This is achieved by:

(a) analysing population processes;
(b) making long-term census counts of natural populations; and
(c) modelling the processes to simulate the natural populations.

(a) Analysing population processes

Three phases can be distinguished in a parasite's life cycle; within each the rate of development and level of mortality must be defined together with causes of mortality.

1. *Free-living development*. The free-living stages of *B. microplus* are the ovipositing females, eggs and non-feeding larvae. Instantaneous development rates as well as the length of the season that is favourable for development need to be defined. New computerized techniques (Dallwitz & Higgins 1978) now allow the use of fluctuating, closely monitored temperatures to replace constant temperature rooms in the

study of development rates. These temperatures can be recorded using microprocessor-based data-loggers (Maywald, O'Neill, Taylor & Baillie, 1985). Development accrues when temperatures are above a critical threshold using the day-degree concept (Baskerville & Emin 1969).

Mortality factors can be classified as either physical or biotic. Climatic factors dominate the tick's life system and, as is evident from the distribution of the tick in northern Australia, it requires warm, wet conditions (Wilkinson 1970). Biotic agents, such as non-specific predatory ants, may cause substantial mortality (Wilkinson 1970) but their activities vary greatly from one pasture to another.

2. *Host-finding*. A parasite's success in finding a host is determined by its median survival time and the instantaneous rate of host parasite contacts (Sutherst, Dallwitz, Utech & Kerr 1978). Tick larvae are sedentary and wait in the pasture for a passing host. Their median survival time of 2–6 weeks and probability of about 0.20 of finding a host enable tick populations to reach pest status on *Bos taurus* cattle at commercial stocking rates as low as 1 beast per 10 ha.

3. *Parasitic phase*. The host-parasite relationship is of paramount importance in determining the survival of both parties. In order for the parasite to survive the host's defences it can reduce its antigenicity (Sprent 1962; Bloom 1979). Reduced virulence is also usual in long established host-parasite relationships (Levin & Pimentel 1981) and the host population may be selected for increased resistance (Fenner 1965). However, as May (1981) has pointed out, the consequences of a particular adaptive strategy for a population may be different from those for an individual. Several evolutionary pathways are successful.

In the case of the cattle tick, the bovine hosts mount an increasingly severe immune response when exposed to larger numbers of larvae (Sutherst & Utech 1981). The response, as measured by tick survival and its frequency distribution in the host population, are the only known tick population stabilizing mechanisms (Sutherst & Utech 1981). As *B. microplus* is a one-host tick, it remains on the same host individual for 3 weeks. During that time it is subjected to the immune response of that individual. The effects of the host's resistance on tick survival are therefore evident as the proportion of ticks which engorge on each animal. This is important because it enables individual hosts with different resistance levels to be identified for management purposes.

In the host-tick relationship, the effects have been defined of tick density, season and the sex, age, breed, nutritional, pregnancy and lactational status of the host on tick mortality rates (Sutherst & Utech 1981).

(b) Making long-term census counts of natural populations

Quantitative data need to be collected on natural populations, firstly to identify major causes of population change for further experimental study. Secondly they provide the basis for estimating damage and benefits from control measures, and thirdly the data enable the testing of population models.

The geographical distribution of parasites needs to be defined by surveys and their abundance and seasonal phenology observed in representative parts of their endemic area. Significant variation in abundance may point to useful mortality factors. Regional censuses are also necessary over a series of years with contrasting weather, simultaneously measuring as many parameters as possible. Quantitative estimates of both the distribution and relative abundance of parasites will help in the use of computerized systems to compare the climatic favourability of different habitats around the globe (Sutherst & Maywald 1985). New epidemiological models (Carter & Prince 1981) are also giving a better description of populations at the edge of their endemic zones.

Data on the frequency distributions as well as the mean values of each parameter (e.g. tick numbers per host or per habitat type) are useful so that any skewness or clumping can be identified and perhaps exploited for control.

Extensive census data on adult cattle ticks have been collected in northern Queensland (Wharton, Harley, Wilkinson, Utech & Kelley 1969; Johnston, Haydock & Leatch 1981), central Queensland (Sutherst, Wharton, Cook, Sutherland & Bourne 1979) and southern Queensland (Mahoney, Wright, Goodger, Mirre, Sutherst & Utech 1981; Sutherst 1983). Interpretation of these data has been most fruitful when the survival rates of free-living and parasitic ticks have also been measured during the observations.

(c) Modelling the processes to simulate natural populations

Modelling the population processes to simulate natural populations is usually a stepwise process. At first, when data are scarce, skeletal models help to orientate the programme and give a feel for the relative importance of difference processes. As data collection progresses the models can be made more realistic.

The use of models in cattle tick control was summarized by Maywald, Dallwitz and Sutherst (1980). A weekly, weather-driven model (TICKI) of tick populations associated with an average host was produced first (Sutherst & Dallwitz 1979). A weekly model, MATIX, was then produced with parameter values fixed to correspond to average and extreme seasons in southern Queensland (Sutherst, Norton & Maywald 1980; Norton, Sutherst & Maywald 1983). Finally a daily model, TICK2, is being developed to describe tick numbers on a herd of cattle in a variable sized pasture with numerous management options. It will summarize our understanding of the ecology of *B. microplus*.

B. Control

The aim of pest management is to manage populations at economically optimal levels. The parasite is only controlled when the cost of doing so is less than the loss that would be incurred if no control was implemented. The ability of each available control method to kill parasites needs to be defined, as well as the cost of labour, management and alternative uses for available resources. In addition, each of the life cycle processes is searched in great detail for new means of interfering with what the parasite is doing to reach the next stage in the life cycle.

The philosophy behind the CSIRO's approach to cattle tick control is to give preference to biological methods of control before integrating other methods which rely on management or chemicals. Reliance on management is minimized because standards are so variable that they prevent progress on an industry-wide scale. Host resistance provides the basis for control but can only be exploited to the extent that is possible without causing adverse side effects (Wharton 1974; Sutherst, Maywald & Sutherland 1980; Sutherst & Utech 1981) and chemical control strategies are designed to delay the emergence of resistance to pesticides (Sutherst & Comins 1979). Likewise, vaccinations are preferred to intensive vector control for suppression of tick-borne diseases (Mahoney 1974; Sutherst & Tahori 1981). In general, our approach can be summarized as "using moderation in all things".

C. Integration of damage and control functions into a management model

Integration of the damage and control relationships, population parameters and economic relationships into a computer model enables us to explore the behaviour of parasite populations and to design control strategies.

As with any other research, it is important to define precisely the objectives of modelling activities. The objectives of modelling cattle tick populations were given by Sutherst and Wharton (1973) and updated by Maywald, Dallwitz and Sutherst (1980); they are shown in table 15.1.

Table 15.1 General objectives of the cattle tick ecology programme

1. To provide a research framework to guide data collection
2. To define the biological relationships in the life cycle
3. To assess the favourability of different geographical areas
4. To understand the population genetics of pesticide resistance and to compare the effects of different control strategies on selection rates
5. To understand the epidemiology of tick-transmitted diseases
6. To design and test integrated control strategies
7. To summarize current ecological knowledge
8. To develop a systems approach to the analysis and management of populations of metazoan parasites of domestic stock
9. To teach students and advisory personnel

We are close to achieving all of those objectives and are now applying the approach to 2- and 3-host ticks, particularly in Africa (Sutherst 1981).

D. Implementation

In the mid-1970s there was concern about the prospects for tick control in southern Queensland. Cattle owners were relying excessively on chemical control and were slow to adopt tick resistant breeds of cattle.

With the knowledge that a tick model was available, the Queensland Department of Primary Industries initiated an intensive regional extension programme on tick control in 1976. A team was established with extension skills in tick control, beef production, economics and research skills in tick ecology. It was encouraged over the years with residential workshops to develop work plans and to evaluate progress. The programme involved the production of a manual on tick control (Powell 1977) and a series of well-planned extension activities. An awareness campaign was mounted to warn of the dangers of total reliance on chemicals, then field days were held to illustrate the advantages of tick resistant cattle. Bus trips were organized to show southern graziers the successes of their northern counterparts with tick resistant cattle. A survey was also conducted into current tick control practices (Elder, Dunwell, Emmerson, Kearnan, Waters, Knott & Morris 1980).

By 1982 tick resistant cattle were predominant in the area and the use of acaricides had declined by 40%. The campaign is a useful example on which to base similar projects in the future.

Buffalo Fly

Superficially there is little resemblance between the buffalo fly and the cattle tick, but from an ecologist's point of view they have many features in common. Fortunately, there is an extensive literature on the closely related horn fly *H. irritans irritans* as well as non-quantitative observations on the buffalo fly.

A. Damage

Damage per fly

The damage caused to cattle by the buffalo fly is very similar to that caused by cattle tick. Relatively low infestations of *Haematobia* spp. have been shown to reduce growth rates of cattle by around 18% (Haufe 1982). The relationship between d and N has not been defined but there is evidence that it is not linear (Kinzer, Houghton, Reeves, Kunz, Wallace & Urquhart 1984). These authors found that 70 flies caused half the weight loss resulting from infestations of 1,300 flies per animal. Both Haufe (1982) and Holroyd, Hirst, Merrifield & Toleman (1984) reported positive correlations between fly numbers and growth rates of cattle. The results above suggest that the irritation caused by flies to more sensitive hosts is more detrimental to the host than is the direct effect of blood loss.

Losses in production are never as severe as those caused by ticks which can kill *Bos taurus* cattle. However, lost production from the 10 million or so cattle involved in northern Australia would amount to at least $50–$100 million p.a. based on published data. Also, unlike cattle tick, both sexes of the fly feed vigorously on the host during adult life. Accurate measurement of losses per fly are hindered by the fly's mobility between individual animals and between adjacent pastures. The movements conceal some of the variation between the fly burdens of individual hosts, so removing the opportunity to use that variation to obtain accurate measurements of losses in production as was possible with cattle tick.

"Buffalo fly lesions" are much more common and severe than those produced by tick bites. They would be prime targets for screw-worm attack which could devastate the cattle industry if *Chrysomyia* is introduced into Australia. The lesions have in the past been attributed to grooming, in response to the irritation from fly feeding (e.g. Roberts 1946). However an unknown species of *Stephanofilaria* has recently been isolated from some of the lesions (Johnson, Parker, Norton, Jaques & Grimshaw 1981). Experimental induction of the lesions is now required to identify the causative agent beyond doubt. We recorded a prevalence of 90% of lesions on a breeding herd in central Queensland in 1982; the distribution of lesions was skewed with most animals having few lesions. The herd was surveyed to investigate the genetics of the hosts' proneness to lesion formation. Areas of the cattle hides affected by the lesions (fig. 15.3) are useless to the leather industry, but fortunately the area of hide affected is small compared to that affected by cattle tick.

Fly numbers and population dynamics

The numbers of flies on cattle vary up to about 3,000 depending upon the age, sex and colour of the animal and the geographical location. In contrast to the larvae of cattle ticks which are non-feeding in the free-living phase, larvae of *H. i. exigua* feed, moult and pupate in the host's dung and it is the adults which emerge to seek a host. The fly's free-living phase is consequently much more involved and is closely tied to the host in contrast to that of the cattle tick.

Immature buffalo flies have to compete for space and food with a rich dung fauna and they also are attacked by non-specific predators and pupal parasitoids. The ecology of the dung pad may be so complex that it prevents as detailed an understanding being obtained of buffalo fly populations as that achieved with cattle tick. Efforts are needed to identify key mortality factors, both physical and biotic, which account for a large proportion of the observed variation in survival of immature buffalo flies. Such an analysis will, hopefully, show that fly populations are affected by only a few "key factors" in dung pads, so rendering the apparent complexity less overwhelming.

The relatively steady maximum size of populations of flies on cattle

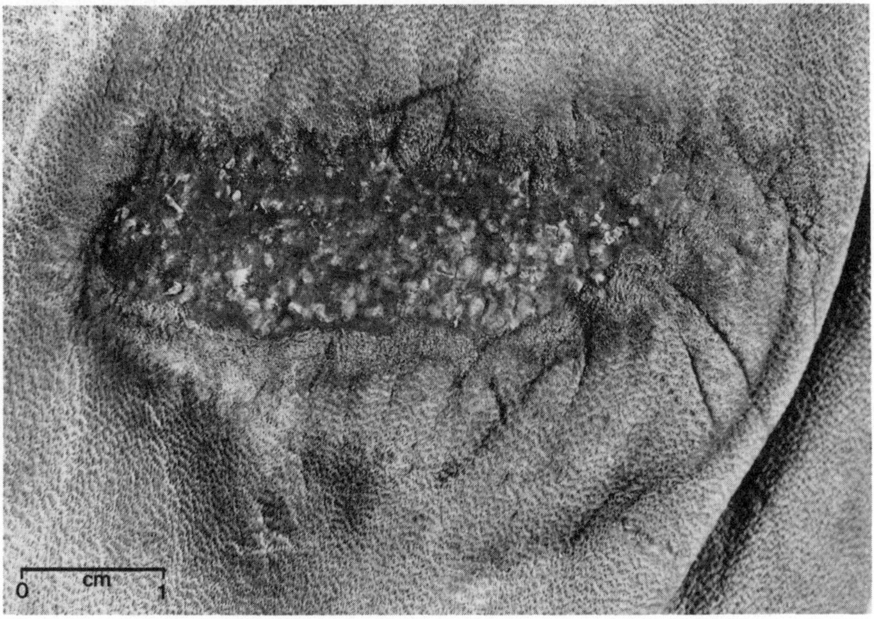

Figure 15.3 Damage to cattle hide associated with feeding by buffalo fly.

each year suggests the presence of a population regulating mechanism. Palmer and Bay (1983) demonstrated density-dependent larval mortality and reductions in pupal weight of *Haematobia* in dung with a range of nutrient values. Those laboratory results need modification to take into account the rapidly declining availability of suitable dung in pads under field conditions. The presence of limiting mortality factors, such as crowding in dung pads or on the host, could explain the disappointing impact of biological control agents on fly burdens seen on cattle. Losses of flies due to new mortality factors may be compensated for by increased survival and fitness of their siblings.

The mobility of the winged adults compensates for their short survival time of about 24 h in the absence of a blood meal. It also enables the fly to reinvade inland areas each summer, following local extinction during severe winters. The probability of finding a host needs to be defined together with the effects of the fly's mobility on chemical control programmes.

Adult *Haematobia* flies have a looser association with individual hosts than do ticks, but they suck blood about every 2 h (Harris & Miller 1969), inducing defensive behaviour and perhaps an immunological response in the host. These responses are likely to cause density dependent reductions in both oviposition and longevity of flies which tend to accumulate on less sensitive individual animals in a herd. The flies oviposit on their host's dung intermittently over a period of days, returning to the same or a different host individual. This transfer from host to host will tend to conceal the effects of differences in defensive reactions of individual hosts on fly numbers, as it does with 3-host ticks (Dicker & Sutherst 1981). Additionally some classes of hosts such as bulls and dark animals are known to be more attractive than others to buffalo flies, but it is not known whether flies feed more successfully on these animals. Successful development to the oviposition stage will depend upon the fly's ability to feed and the rate of oogenesis relative to the longevity of the fly on the host. These parameters will have to be investigated in relation to the favourability of different hosts, to fly densities on the host and to climatic factors.

We plan to obtain census data on buffalo fly populations in each infested climatic zone in Australia early on in our programme, to provide a basis for our ecological studies. In order to obtain quantitative data we have developed a trap with which to remove flies temporarily from cattle for counting. The trap will also be a useful experimental tool to study numerous aspects of fly biology including dispersal, effects of flies on cattle and host-seeking behaviour.

B. Control

The detrimental effects of low numbers of *Haematobia* spp. on cattle indicate the need for any successful control programme to be extremely efficient. In fact the possibility of eradication deserves serious consideration as the ultimate goal of research into this pest. Any programme which is to be so highly efficient will need to combine different approaches to reducing fly numbers.

In Australia, the greatest research effort has been aimed at control of buffalo fly using an exotic dung fauna, but results have so far been disappointing. New approaches to chemical control using slow-release technology are being developed with promising results, but they will need more careful use than we have seen in the past if they are to make a contribution before resistance develops. We plan to explore the possibility of vaccination using non-salivary antigens, as suggested by observations of Schlein and Lewis (1976) on *Stomoxys calcitrans* and Nogge and Giannetti (1980) on tsetse fly.

C. Integration

The objectives of our buffalo fly programme are comparable to items 1–4, 6, 7 and 9 in table 15.1. In addition the programme is aimed specifically at helping in the assessment of the potential of biological control agents in preventing economic losses.

The cattle tick modelling was carried out at a time when few ecologists had much experience in modelling. Experience since that time will greatly assist with modelling buffalo fly populations. As a start we have developed a climatic matching programme which compares general favourability of different places and seasons for any poikilothermic animals. Geographical comparisons can be made of the values of a population growth index and of climatic stress indices on a weekly or annual basis. Our initial experience in making global comparisons of the climatic favourability for *Haematobia* and for *B. microplus* have been most illuminating. The programme has the potential for wide application in parasitology and biological control.

Discussion

One of the most damaging criticisms of modern biological research is that there is too much "ad hocery". This criticism is even more relevant in applied research where specific problems can be defined and objectives set. It is in such situations that systems techniques are so valuable.

Some ecological attributes of the cattle tick and buffalo fly are compared in table 15.2. By identifying the similarities we can use our tick experience to expedite the buffalo fly research programme.

We find that most population processes are common to both and indeed to all parasites. This enables us to specify the particular data that we require. The differences centre around how to measure the processes and here we need different techniques to suit each parasite. Most importantly our goals are defined so that the programme has direction.

The degree of similarity in the population processes of these two superficially different parasites is encouraging to ecologists looking for efficient research strategies with which to analyse parasite problems. Other major parasite groups such as helminths can be fitted into the same overall framework. A bonus from such a system is that many of the computer programs developed to describe a life cycle process for a specific parasite can be readily adapted for a new study thus reducing expensive programming costs.

The problem of priorities has to be resolved in the light of what is most desirable — i.e. the major gaps or sensitive parameters — and what is practicable. We intend to build a simplified population model as soon as possible to explore the sensitivity of various aspects of the life system. The physical environment in dung pads will be intensively modelled because it provides the driving force for development and mortality of all dung fauna. Concurrently we will measure the age specific mortality of fly eggs, larvae and pupae and relate it to their physical environment, to fly density and to the activity of both native and exotic dung fauna.

The long-term objective of parasitologists should be to design integrated programmes which simultaneously manage all parasites causing economic damage to domestic animals. Tick resistant cattle reduced the tick problem

Table 15.2 Comparison of knowledge of ecological and other characteristics of *B. microplus* and *H. exigua*

SIMILARITIES		
Place of origin	Indonesia	
Geographical distribution in Australia	Wet tropics and subtropics	
Host	Bovidae	
Damage	liveweight loss milk yield reduction hide disfigurement lesions — screw worm fly	
Type of parasite	ectoparasite, direct life cycle.	
Evolutionary strategy	mostly *r* selected	
Population processes	free-living development host finding dispersal parasitic feeding and mating oviposition	
Control	chemicals	
Sampling immature stages	difficult and laborious	
Culture	routine	
DIFFERENCES		
	B. microplus	*H. exigua*
Damage	transmit piroplasms (*Babesia; Anaplasma*)	? transmit *Stephanofilaria* sp.
Free-living phase	larvae have a non-feeding phase	larvae feed, moult and pupate off host
Host finding	larvae host seeking; long lived; sessile	adults host seeking; short-lived; mobile
Parasitic phase	continuous attachment to same host affected by host resistance	frequent intermittent feeding on same or different host host resistance unknown; host attractiveness variable
Control	host resistance pasture rotation possible	fly mobility enhances chemical control within herds; hinders single property programmes
Sampling adults	accurate visual counts of ticks of narrow age class	visual counts approximate, subject to bias; accurate trapping counts age determination laborious

but aggravated the buffalo fly problem by reducing the use of chemicals. In the long run this development should be advantageous in extending the useful life of chemicals. In the meantime increased understanding of the ecology of the buffalo fly is necessary.

Acknowledgments

Mr J. Green photographed the lesion on the cattle hide and Dr R.D. Hughes, in particular, made constructive and helpful criticism of the manuscript. The AMRC contributed funding to the tick and buffalo fly projects. As always Mrs D. Dunn and Barbara Watts cheerfully typed the manuscript and Ms M. Krohn gave assistance in the library.

References

Baskerville, G.L. and Emin, P. 1969. Rapid estimation of heat accumulation from maximum and minimum temperatures. *Ecology* **50**:514–17.

Bloom, B.R. 1979. Games parasites play: How parasites evade immune surveillance. *Nature (London)* **279**:21–26.

Carter, R.N. and Prince, S.D. 1981. Epidemic models used to explain biogeographical distribution limits. *Nature (London)* **293**:644–45.

Clark, L.R., Geier, P.W., Hughes, R.D. and Morris, R.F. 1967. *The Ecology of insect populations in theory and practice*. London: Methuen.

Dallwitz, M.J. and Higgins, J.P. 1978. *User's guide to DEVAR. A computer program for estimating development rate as a function of temperature*. CSIRO Division of Entomology Report No. 2.

Dicker, R.W. and Sutherst, R.W. 1981. Control of the bush tick (*Haemaphysalis longicornis*) with Zebu × European cattle. *Australian Veterinary Journal* **57**:66–68.

Elder, J.K., Dunwell, G.H., Emmerson, F.R., Kearnan, J.F., Waters, K.S., Knott, S.G. and Morris, R.S. 1980. A survey concerning cattle tick control in Queensland. 1. Producer attitudes. In *Ticks and tick-borne diseases*, ed. L.A.Y. Johnston and M.G. Cooper, 35–39. Sydney: Australian Veterinary Association.

Fenner, F. 1965. Myxoma virus and *Oryctolagus cuniculus*: Two colonizing species. In *The genetics of colonizing species*, ed. H.G. Baker and G.L. Stebbins, 485–99. New York: Academic Press.

Harris, R.L. and Miller, S.A. 1969. A technique for studying the feeding habits of the horn fly. *Journal of Economic Entomology* **62**:279–80.

Haufe, W.O. 1982. Growth of range cattle protected from horn flies (*Haematobia irritans*) by ear tags impregnated with fenvalerate. *Canadian Journal of Animal Science* **62**:567–73.

Holling, C.S. 1964. The analysis of complex population processes. *Canadian Entomologist* **96**:335–47.

Holroyd, R.G., Hirst, D.J., Merrifield, A.W. and Toleman, M.A. 1984. The effect of spraying for buffalo fly (*Haematobia irritans exigua*) on infestations, growth rate and lesion development on *Bos indicus* × *B. taurus* cattle in the dry tropics of north Queensland. *Australian Journal of Agricultural Research* **3**:596–608.

James, A.D. and Ellis, P.R. 1980. The evaluation of production and economic effects of disease. *Proceedings of the 2nd International Symposium, Veterinary Epidemiology and Economics, 1979* 363–72.

Johnston, L.A.Y., Haydock, K.P. and Leatch, G. 1981. The effect of two systems of cattle tick (*Boophilus microplus*) control on tick populations, transmission of *Babesia* spp. and *Anaplasma* spp. and production of Brahman crossbred cattle in the dry tropics. *Australian Journal of Experimental Agricultural and Animal Husbandry* **21**:256–67.

Johnson, S.J., Parker, R.J., Norton, J.H., Jaques, P.A. and Grimshaw, A.A. 1981. Stephanofilariasis in cattle. *Australian Veterinary Journal* **57**:411–13.

Kaiser, M.N., Sutherst, R.W. and Bourne, A.S. 1982. Relationship between ticks and Zebu cattle in southern Uganda. *Tropical Animal Health and Production* **14**:63–74.

Kinzer, H.G., Houghton, W.E., Reeves, J.M., Kunz, S.E., Wallace, J.D. and Urquart, N.S. 1984. Influence of horn flies on weight loss in cattle with notes on prevention of loss by insecticide treatment. *Southwestern Entomologist* **9**:212–17.

Levin, S.A. and Pimentel, D. 1981. Selection of intermediate rates of increase in parasite-host systems. *American Naturalist* **117**:308–15.

Mahoney, D.F. 1974. The application of epizootiological principles in the control of babesiosis in cattle. *Bulletin de l'Office International des Epizooties* **81**:123–38.

Mahoney, D.F., Wright, I.G., Goodger, B.V., Mirre, G.B., Sutherst, R.W. and Utech, K.B.W. 1981. The transmission of *Babesia bovis* in herds of European and Zebu × European cattle infested with the tick, *Boophilus microplus. Australian Veterinary Journal* **57**:461–69.

May, R.M. 1981. Population biology of parasitic infections. In *The current status and future of parasitology*, ed. K.S. Warren and E.F. Purcell, 208–35. New York: Josiah Macy Jr. Foundation.

Maywald, G.F., O'Neill, B.M., Taylor, M.F. and Baillie, S.A. (1985). A low-cost microprocessor-based field data logger. *CSIRO Division of Entomology Report No. 36.*

Maywald, G.F., Dallwitz, M.J. and Sutherst, R.W. 1980. A systems approach to cattle tick control. *Proceedings 4th Biennial Conference Simulation Society of Australia*. 132–39.

Mellor, W., O'Rourke, P.K. and Waters, K.S. 1983. Tick infestations and their effects on growth of *Bos indicus* and *Bos taurus* cattle in the wet tropics. *Australian Journal of Experimental Agriculture and Animal Husbandry* **23**:348–53.

Nogge, G. and Giannetti, M. 1980. Specific antibodies: A potential insecticide. *Nature (London)* **209**:1028–29.

Norton, G.A., Sutherst, R.W. and Maywald, G.F. 1983. A framework for integrating control methods against the cattle tick, *Boophilus microplus* in Australia. *Journal of Applied Ecology* **20**:489–505.

Palmer, W.A. and Bay, D.E. 1983. Effects of intraspecific competition and nitrogen content of manure on pupal weight, survival and reproductive potential of the horn fly, *Haematobia irritans irritans* (L.). *Protection Ecology* **5**:153–60.

Powell, R.T. 1977. Project tick control. *Queensland Agricultural Journal* **103**:443–74.

Roberts, F.H.S. 1946. The buffalo fly. *Queensland Agricultural Journal* **63**:112–16.

Schlein, Y. and Lewis, C.T. 1976. Lesions in haematophagous flies after feeding on rabbits immunized with fly tissues. *Physiological Entomology* **1**:55–59.

Southwood, T.R.E. and Norton, G.A. 1973. Economic aspects of pest management strategies and decisions. In *Insects: Studies in population management*, ed. P.W. Geier, L.R. Clark, D.J. Anderson, and H.A. Nix, 168–84. Canberra: Ecological Society of Australia.

Sprent, J.F.A. 1962. Parasitism, immunity and evolution. In *The evolution of living organisms*, ed. G.W. Leeper 149–65. Melbourne: Melbourne University Press.

Sutherst, R.W. 1981. Is the Australian pest management approach to tick control relevant to Africa? In *Tick biology and control*, ed. G.B. Whitehead and J.D. Gibson 179–85. Grahamstown: Rhodes University.

Sutherst, R.W. 1983. Variation in the numbers of cattle tick, *Boophilus microplus*, in a moist habitat made marginal by low temperatures. *Journal of the Australian Entomological Society* **22**:1–5.

Sutherst, R.W. and Comins, H.N. 1979. The management of acaricide resistance in the cattle tick, *Boophilus microplus* (Canestrini) (Acari:Ixodidae), in Australia. *Bulletin of Entomological Research* **69**:519–40.

Sutherst, R.W. and Dallwitz, M.J. 1979. Progress in the development of a population model for the cattle tick *Boophilus microplus. Proceedings of the 4th International Congress of Acarology, 1974* 557–63.

Sutherst, R.W., Dallwitz, M.J., Utech, K.B.W. and Kerr, J.D. 1978. Aspects of host finding by the cattle tick *Boophilus microplus. Australian Journal of Zoology* **26**:159–74.

Sutherst, R.W. and Maywald, G.F. 1985. A computerised system for matching climates in ecology. *Agriculture, Ecosystems and Environment* **13**:281–299.

Sutherst, R.W., Maywald, G.F. and Sutherland, I.D. 1980. The value of host resistance against cattle tick (*Boophilus microplus*) in Australia. *Proceedings of the 2nd International Symposium on Veterinary Epidemiology and Economics* 1979, 408–15.

Sutherst, R.W., Maywald, G.F., Kerry, J.D. & Stegman, D.A. 1983. The effect of cattle tick (*Boophilus microplus*) on the growth of *Bos indicus* x *B. taurus* steers. *Australian Journal of Agricultural Research* **34**:317–27.

Sutherst, R.W., Norton, G.A., Barlow, N.D., Conway, G.R., Birley, M. and Comins, H.N. 1979. An analysis of management strategies for cattle tick (*Boophilus microplus*) control in Australia. *Journal of Applied Ecology* **16**:359–82.

Sutherst, R.W., Norton, G.A. and Maywald, G.F. 1980. Analysis of control strategies for cattle tick on Zebu × British cattle. In *Ticks and tick-borne diseases*, ed. L.A.Y. Johnston and M.G. Cooper, 46–51. Sydney: Australian Veterinary Association.

Sutherst, R.W. and Tahori, A.S. 1981. Vector control: Appraisal and future perspectives. In *Advances in the control of Theileriosis*, ed. A.D. Irvin, M.P. Cunningham and A.S. Young, 173–76. The Hague: Nijhoff.

Sutherst, R.W. and Utech, K.B.W. 1981. Controlling livestock parasites with host resistance. In *CRC handbook of pest management in agriculture*, Vol. II. ed. D. Pimentel, 385–407. Boca Raton, Fla.: CRC Press.

Sutherst, R.W. and Wharton, R.H. 1973. Preliminary considerations of a population model for *Boophilus microplus* in Australia. *Proceedings of the 3rd International Congress of Acarology, 1971*. 797–801.

Sutherst, R.W., Wharton, R.H., Cook, I.M., Sutherland, I.D. and Bourne, A.S. 1979. Long-term population studies on the cattle tick (*Boophilus microplus*) on untreated cattle selected for different levels of tick resistance. *Australian Journal of Agricultural Research* **30**:353–68.

Sutherst, R.W., Wharton, R.H. and Utech, K.B.W. 1978. Guide to studies on tick ecology. *CSIRO Division of Entomology Technical Paper No. 14*.

Utech, K.B.W., Wharton, R.H. and Kerr, J.D. 1978. Resistance to *Boophilus microplus* (Canestrini) in different breeds of cattle. *Australian Journal of Agricultural Research* **29**:885–95.

Watt, K.E.F. 1968. *Ecology and resource management*. New York: McGraw Hill.

Wharton, R.H. 1974. Ticks with special emphasis on *Boophilus microplus*. In *Control of arthropods*, ed. R. Pal and R.H. Warton, 35–52.

Wharton, R.H., Harley, K.L.S., Wilkinson, P.R., Utech, K.B.W. & Kelley, B.M.

1969. A comparison of cattle tick control by pasture spelling, planned dipping, and tick-resistant cattle. *Australian Journal of Agricultural Research* **20**:783–97.

Wharton, R.H. & Roulston, W.J. 1970. Resistance of ticks to chemicals. *Annual Review of Entomology* **15**:381–404.

Wharton, R.H., Utech, K.B.W. & Sutherst, R.W. 1973. Tick resistant cattle for the control of *Boophilus microplus. Proceedings of the 3rd International Congress of Acarology, 1971*:697–700.

Wilkinson, P.R. 1970. Factors affecting the distribution and abundance of the cattle tick in Australia: Observations and hypotheses. *Acarologia* **12**:492–508.

List of Contributors

Allen, John R. Dr
Department of Veterinary Microbiology,
Western College of Veterinary Medicine,
University of Saskatchewan,
Saskatoon, Saskatchewan, Canada
Ballantyne, Robert J. Dr
School of Applied Science,
Riverina-Murray Institute of Higher Education,
Wagga Wagga, New South Wales, Australia
Cannon, Lester R.G. Dr
Queensland Museum,
Queensland Cultural Centre,
Brisbane, Queensland, Australia
Dobson, Colin Prof.
Department of Parasitology,
University of Queensland,
St Lucia, Queensland, Australia
Heath, Allen C.G. Dr
Wallaceville Animal Research Centre,
Research Division,
Ministry of Agriculture and Fisheries,
Upper Hutt, New Zealand
Hobbs, R.P.
Department of Zoology,
University of New England,
Armidale, New South Wales, Australia
Lester, Robert J.G. Dr
Department of Parasitology,
University of Queensland,
St Lucia, Queensland, Australia
Mahoney, David F. Dr
CSIRO Division of Tropical Animal Science,
Long Pocket Laboratories,
Indooroopilly, Queensland, Australia
Nutting, William B. Prof.
Zoology Department,
University of Massachusetts,
Amherst, Massachusetts, USA

Pearson, John C. Dr
Department of Parasitology,
University of Queensland,
St Lucia, Queensland, Australia
Rickard, Michael D. Dr
Department of Paraclinical Sciences,
Veterinary Clinical Centre,
University of Melbourne,
Princes Highway,
Werribee, Victoria, Australia
Rohde, Klaus Dr
Department of Zoology,
University of New England,
Armidale, New South Wales, Australia
Rzepczyk, Christine M. Dr
Queensland Institute of Medical Research,
Herston, Queensland, Australia
Sutherst, Robert W. Dr
CSIRO Division of Entomology,
Long Pocket Laboratories,
Indooroopilly, Queensland, Australia
Weilgama, Danister J. Dr
Veterinary Research Institute,
Peradeniya, Sri Lanka
Wright, Ian G. Dr
CSIRO Division of Tropical Animal Science,
Long Pocket Laboratories,
Indooroopilly, Queensland, Australia
Yong Weng Kwong, Dr
Regional Veterinary Laboratory,
Division of Animal Health,
Department of Agriculture,
Hamilton, Victoria, Australia

Biography of John Frederick Adrian Sprent, CBE, FAA

The following quote from the citation for the Mueller Medal of the Australian and New Zealand Society for the Advancement of Science, which he received in 1981, summarizes John Sprent's contribution to education and science.

> From a small beginning as a unit in the Faculty of Veterinary Science in the University of Queensland with John Sprent as the sole staff member the Department of Parasitology has been built under his leadership into a dynamic centre of research that is world-renowned. At first John Sprent's research was devoted to the parasites of domestic animals, and especially to their public health significance. But when the immediate needs for the study of domestic animals' parasites had been satisfied he turned his attention to the parasites of the native Australian fauna which he reasoned might well harbour unique parasites due to their isolation from fauna elsewhere in the world. He discovered interesting life history patterns in several new nematode species he found in Australian native animals and especially fascinating ones resident in carpet snakes. This has led to a study of snake parasites in all the continents of the world.
>
> John Sprent's studies lead him to write *Parasitism* [1963], a book which became recognised internationally as a landmark in immunology. He has continued to the present day to publish research papers on nematode and helminth life histories and on the immunological, allergic and tolerance reactions of their hosts, having produced almost 100 research papers on these topics in his nigh on thirty years at the University of Queensland.

John Sprent has developed his research interests across many important areas of parasitology including the immunology and pathology of the infected host as well as his detailed, classical studies on taxonomy, life cycles, zoogeography and evolution of parasites, particularly ascaridoid nematodes. The depth and breadth of his scientific perception is illustrated in his paper "Parasitism, Immunity and Evolution" published in 1959 in *The Evolution of Living Organisms* (ed. G.N. Leeper, Melbourne: Melbourne University Press). This work has had a major impact on the development of our understanding of host-parasite relationships over the past twenty years, and has greatly influenced the direction of research on parasitic diseases of veterinary and medical importance. His further work on parasites themselves has helped to clarify many difficult problems associated with the transmission of parasitic zoonoses to man and domestic animals. The following gives an account of John Sprent's career and the many academic accolades he has received for his work in education and his research in science.

John Sprent was born in London in 1915, educated at Shrewsbury School and graduated MRCVS with the Colman Silver Medal in Veterinary Medicine and the NVMA Gold Medal in Pathology from the Royal College of Veterinary Surgeons 1939, and B.Sc., First Class Honours from University of London, 1943. He obtained his Ph.D. from the University of London in 1945 while serving as a veterinary research officer in Nigeria.

Between 1945 and 1952 he held four research fellowships and worked variously in the Ministry of Agriculture, England, Chicago and Ontario before arriving in Australia in 1952 as senior lecturer in Veterinary Parasitology, University of Queensland. His excellence as a research worker and educator was recognized early with the award of D.Sc. London 1953 and his rapid promotion to Research Professor of Parasitology 1954 and Professor of Parasitology 1956.

John Sprent has given long and continuing outstanding service to the University of Queensland, the veterinary profession and primary industry, to the state of Queensland in particular and as a result to Australia as a whole. His achievements in these areas are a measure of his great industry, scholarship and humanity. They are a signal for all who seek to serve with merit and excellence.

John Sprent was instrumental in the development of parasitology as a discipline and centre of excellence in research within the university. He established the Department of Parasitology in 1960 at a time when he was occupied as dean of the Faculty of Veterinary Science 1960–63. At this time he was moving to form the Australian Society for Parasitology of which he was elected foundation president 1964. By 1964 the Department of Parasitology, as it presently is structured, had been developed with the appointment of six parasitologists.

It is not surprising that John Sprent quickly accrued academic honours for his work. He was appointed Fellow of the Royal College of Veterinry Surgeons in 1960 for his achievements in veterinary science. In 1961 he was elected to the New York Academy of Sciences for his research and awarded the Payne Exhibition, University of Melbourne, for research in parasitology. This early period of recognition culminated in the prestigious award of the Henry Baldwin Ward Medal from the American Society of Parasitologists (1962) and his election as Fellow of the Australian Academy of Science (1964). At this time he was also elected president, Section L, Veterinary Science, ANZAAS (1962).

Having established a burgeoning school of research and teaching in parasitology which serviced the requirements of four faculties in the university, John Sprent willingly gave his services internationally. He has served the World Health Organization as a consultant and member of their expert panel on parasitic diseases from 1963 to date. He has travelled extensively in pursuit of his research and teaching through Europe, the

Americas, Africa, India, and the Middle and Far East and as a consequence has brought international recognition that the Department of Parasitology in Queensland is a reputable centre of teaching and research of agricultural, veterinary, medical and scientific importance.

The high regard of the international scientific community for him obtains today and has grown with time. He was elected a Foundation Fellow of the Australian College of Veterinary Scientists (1971), became a member of the CSIRO Postgraduate Studentships Selection Committee (1971–72), and president, Section 16, ANZAAS (1972). He was elected Fellow of the Australian Society for Parasitology (1973) and council member, Australian Academy of Science 1974–77.

It is typical of men of great ability that their industry is not spent selfishly. John Sprent has continuously accepted the rôle of administration as a proper adjunct to his research and educational responsibilities. This was seen early in his career here by his acceptance of the position of dean of Veterinary Science 1960–63 at a time when the Department of Parasitology was still in its infancy. Since that time John Sprent has been charged with many tasks of responsibility by this university. He was chairman, Research Committee of the Senate, a member of the Professorial Board as head of the Department of Parasitology and later by election to the Academic Board. He has served on four faculty boards with distinction and more recently has frequently been used by the vice-chancellor to serve on many important committees. His service to education has recently been extended internationally with his appointment as external examiner, University of Malaya in 1981.

John Sprent reaches his 71st birthday this year yet he still continues his work at a pace that would tire many younger men. He has been editor-in-chief of the International Journal for Parasitology since 1974, work for which he has been highly commended by the Australian Society for Parasitology at its annual general meeting each year. At the same time he continues to research and publish his work.

It is not difficult to outline the attributes of a man with the stature of John Sprent. He has a long and illustrious career in science and education within the state of Queensland. Many hundreds of younger men and women bear his hallmark of merit and through the skills imparted by him they responsibly service the requirement of our community for excellence in agriculture, veterinary science, medicine, science and education within Queensland, and throughout Australia and the world. From a small beginning his efforts have ensured that the Department of Parasitology has grown to play a key rôle in the development of postgraduate education in this university. The department has produced more than fifty successful doctoral graduates, many of whom now hold important positions in the Commonwealth Scientific and Industrial Research Organization (CSIRO), departments of agriculture, universities, colleges of advanced education and institutes of technology. All these men and women would speak of John Sprent as a great man, first for his helpfulness, humanity and sensitivity, and second as an outstanding scholar and scientist.

Now that he is retired it is fitting and timely that he should be rewarded by international organizations for his long meritorious career in the University of Queensland in particular, and science in general. In 1979 he was granted life membership of the Australian Veterinary Association and elected to honorary membership of the Helminthological Society of Washington (USA). In 1980 he was elected Honorary Member, American Society of Parasitologists, and in 1981 he was awarded the Mueller Medal by ANZAAS and elected Fellow of the Queensland Institute of Medical Research. John Sprent was appointed Commander of the Most Excellent Order of the British Empire in 1985.

Publications of J.F.A. Sprent

Sprent, J.F.A. 1946a. Critical anthelmintic tests in cattle. *Veterinary Journal* **102**:83–87.

_______. 1946b. Some observations on a disease of Zebu cattle associated with infestation by the hookworm, *Bunostomum phlebotomum*. *Journal of Comparative Pathology and Therapeutics* **56**:149–59.

_______. 1946c. Some observations on the bionomics of *Bunostomum phlebotomum*, a hookworm of cattle. *Parasitology* **37**:202–10.

_______. 1946d. Some observations on the incidence of bovine helminths in Plateau Province, Northern Nigeria. *Veterinary Journal* **102**:36–40.

_______. 1946e. Studies on the life-history of *Bunostomum phlebotomum* (Railliet, 1900), a hookworm parasite of cattle. *Parasitology* **37**:192–201.

_______. 1946f. Immunological phenomena in the calf, following experimental infection with *Bunostomum phlebotomum*. *Journal of Comparative Pathology and Therapeutics* **56**:286–97.

_______. 1946g. Some observations relating to the critical anthelmintic test. *Veterinary Record* **58**:487–88.

_______. 1947. Resistance of animals to worm infestations. 1. *Report Animal Health Trust* (1946–47), p.12.

Sprent, J.F.A. and Chen, H.H. 1949a. Immunological studies in mice infected with the larvae of *Ascaris lumbricoides*. I. Criteria of immunity and immunizing effect of isolated worm tissues. *Journal of Infectious Diseases* **84**:111–24.

Sprent, J.F.A. 1949b. On the toxic and allergic manifestations produced by the tissues and fluids of Ascaris. I. Effect of different tissues. *Journal of Infectious Diseases* **84**:221–29.

_______. 1950a. On the toxic and allergic manifestations caused by the tissues and fluids of Ascaris. II. Effect of different chemical fractions on worm-free, infected and sensitized guinea pigs. *Journal of Infectious Diseases* **86**:146–58.

_______. 1950b. Observations on the life history of *Ascaris columnaris*. *Journal of Parasitology* **36**:suppl. p.29.

_______. 1951a. Observations on the migratory activity of the larvae of *Toxascaris transfuga* (Rud. 1819) Baylis and Daubney 1922. *Journal of Parasitology* **37**:326–27.

_______. 1951b. On the toxic and allergic manifestations caused by the tissues and fluids of Ascaris. III. Hypersensitivity through infection in the guinea pig. *Journal of Infectious Diseases* **88**:168–77.

_______. 1951c. On the migratory behaviour of the larvae of various *Ascaris* species in mice. *Journal of Parasitology* **37**:suppl. p.21.

Labzoffsky, N.A. and Sprent, J.F.A. 1952a. Tularemia among beaver and muskrat in Ontario. *Canadian Journal of Medical Sciences* **30**:250–55.

Sprent, J.F.A. 1952b. Anatomical distinction between human and pig strains of *Ascaris. Nature (London)* **170**:627–28.

_______. 1952c. (1) Migratory behaviour of *Ascaris* larvae in mice. (2) The dentigerous ridges of the human and pig *Ascaris. Transactions of the Royal Society of Tropical Medicine and Hygiene* **46**:378.

_______. 1952d. On an ascaris parasite of the fisher and marten, *Ascaris devosi* sp. nov. *Proceedings of the Helminthological Society of Washington* **19**:27–37.

_______. 1952e. On the migratory behaviour of the larvae of various *Ascaris* species in white mice. I. Distribution of larvae in tissues. *Journal of Infectious Diseases* **90**:165–76.

_______. 1953a. Intermediate hosts in *Ascaris* infections. *Journal of Parasitology* **39**:suppl. p.38.

_______. 1953b. On the life history of *Ascaris devosi* and its development in the white mouse and the domestic ferret. *Parasitology* **42**:244–58.

_______. 1953c. On the migratory behaviour of the larvae of various *Ascaris* species in white mice. II. Longevity of encapsulated larvae and their resistance to freezing and putrefaction. *Journal of Infectious Diseases* **92**:114–17.

_______. 1954. The life cycles of nematodes in the family Ascarididae Blanchard 1896. *Journal of Parasitology* **40**:608–17.

_______. 1955a. On the invasion of the central nervous system by nematodes. I. The incidence and pathological significance of nematodes in the central nervous system. *Parasitology* **45**:31–40.

_______. 1955b. On the invasion of the central nervous system by nematodes. II. Invasion of the nervous system in ascariasis. *Parasitology* **45**:41–55.

_______. 1955c. The life history of *Ophidascaris filaria* in the carpet snake (*Morelia argus*). *Journal of Parasitology* **41**:suppl. p.40.

_______. *1955d. The life history of Toxocara cati* in the cat. *Journal of of Parasitology* **41**:suppl. p. 40.

_______. 1955e. The life history of *Toxascaris leonina* in the cat. *Journal of Parasitology* **41**:suppl. pp. 39-40.

_______. 1956. The life history and development of *Toxocara cati* (Schrank 1788) in the domestic cat. *Parasitology* **46**:54–78.

_______. 1957a. The development of *Toxocara canis* (Werner 1782) in the dog. *Journal of Parasitology* **43**:suppl. p.45.

_______. 1957b. A new species of *Neoascaris* from *Rattus assimilis*, with a redefinition of the genus. *Parasitology* **47**:350–60.

———. 1958a. Observations on the development of *Toxocara canis* (Werner 1782) in the dog. *Parasitology* **48**:184–209.

Sprent, J.F.A. and English, P.B. 1958b. The large roundworms of dogs and cats — A public health problem. *Australian Veterinary Journal* **34**:161–71.

Sprent, J.F.A. 1959a. The life history and development of *Toxascaris leonina* (von Linstow 1902) in the dog and cat. *Parasitology* **49**:330–71.

———. 1959b. Observations on the development of ascaridoid nematodes of the carpet snake. *Journal of Parasitology* **45**:suppl. p.35.

Sprent, J.F.A. and Mines, J.J. 1960a. A new species of *Amplicaecum* (Nematoda) from the carpet snake (*Morelia argus variegatus*): With a re-definition and a key for the genus. *Parasitology* **50**:183–98.

Sprent, J.F.A. 1961a. Post-parturient infection of the bitch with *Toxocara canis. Journal of Parasitology* **47**:284.

Sprent, J.F.A., Hoyte, H.M.D. and Pearson, J.C. 1961b. *Notes on methods used in parasitology, with a bibliography and keys for identification of parasites of domestic animals*. Appendices A–E. St Lucia: University of Queensland Press. 84pp.

Timourian, H., Dobson, C. and Sprent, J. F. A. 1961c. Precipitating antibodies in the carpet snake against parasitic nematodes. *Nature* **192**:996–97.

Sprent, J.F.A. 1962a. Parasitism, immunity and evolution. In *The evolution of living organisms*, ed. G.W. Leeper, 149–65. Parkville, Victoria: Melbourne University Press.

———. 1962b. The evolution of the Ascaridoidea. *Journal of Parasitology* **48**:818–24.

———. 1963a. Visceral larva migrans — Presidential address. A.N.Z.A.A.S. Section L. 1962. *Australian Journal of Science* **25**:344–54.

———. 1963b. *Parasitism — An introduction to parasitology and immunology for students of biology, veterinary science, and medicine.* St Lucia: University of Queensland Press. 92pp.

———. 1963c. The life history and development of *Amplicaecum robertsi*, an ascaridoid nematode of the carpet python (*Morelia spilotes variegatus*). I. Morphology and functional significance of larval stages. *Parasitology* **53**:7–38.

———. 1963d. The life history and development of *Amplicaecum robertsi*, an ascaridoid nematode of the carpet python (*Morelia spilotes variegatus*). II. Growth and host specificity of larval stages in relation to the food chain. *Parasitology* **53**:321–37.

———. 1963e. Tropical parasitology. *Nature (London)* **197**:1256–57.

Sprent, J.F.A., Hogarth Scott, R.S. and Timourian, H. 1963f. Diffusion chambers for the collection of helminth antigens. *Nature (London)* **200**:913.

Sprent, J.F.A. 1964a. A study of adaptation tolerance — The growth of ascaridoid larvae in indigenous and non-indigenous intermediate hosts. *Proceedings, First UNESCO Regional Symposium on Scientific Knowledge of Tropical Parasites* University of Singapore, 5-9 November 1962, 261–66.

Sprent, J.F.A.and Barrett, Marjorie G. 1964b. Large roundworms of dogs and cats: Differentiation of *Toxocara canis* and *Toxascaris leonina. Australian Veterinary Journal* **40**:166–71.

English, P.B. and Sprent, J.F.A. 1965a. The large roundworms of dogs and cats: Effectiveness of piperazine salts against immature *Toxocara canis* in prenatally infected puppies. *Australian Veterinary Journal* **41**:50–53.

Sprent, J.F.A. 1965b. Ascaridoid larva migrans: Differentiation of larvae in tissues. *Transactions of the Royal Society of Tropical Medicine and Hygiene*, **59**:365–66.

———. 1966a. The components of host specificity in infections with ascaridoid nematodes. *Proceedings, First International Congress of Parasitology* Rome, September 1964, pp.15–18.

———. 1966b. Observations relating to the diagnosis of visceral larva migrans. *Proceedings, First International Congress of Parasitology* Rome, September 1964, 802–3.

———. 1966c. The Australian School of Parasitology. *Australian Journal of Science* **29**:40–44.

Sprent, J.F.A. and McKeown, Ann. 1967a. A study on adaptation tolerance: Growth of ascaridoid larvae in indigenous and non-indigenous hosts. *Parasitology* **57**:549–54.

Sprent, J.F.A., Hoyte, H.M.D., Pearson, J.C. and Waddell, A.H. 1967b. *Notes on methods used in parasitology*. 2nd ed., St. Lucia: University of Queensland Press. 102pp.

Sprent, J.F.A. 1968a. Nematode larva migrans. *Australian Veterinary Association, New South Wales Veterinary Proceedings* **4**:18–19.

———. 1968b. Notes on *Ascaris* and *Toxascaris*, with a definition of *Baylisascaris* gen.nov. *Parasitology* **58**:185–98.

———. 1968c. *Ophidascaris* and *Lagochilascaris*: A comparison. *Proceedings, 8th International Congresses on Tropical Medicine and Malaria* Teheran, 7–15 September 1968. 191–92.

———. 1969a. Studies on ascaridoid nematodes in pythons: Redefinition of *Ophidascaris filaria* and *Polydelphis anoura. Parasitology* **59**:129–40.

———. 1969b. Nematode larva migrans. *New Zealand Veterinary Journal* **17**:39–48.

———.1969c. Evolutionary aspects of immunity in zooparasitic infections. In *Immunity to parasitic animals*, ed. G.J. Jackson, R. Herman, and I. Singer, vol. 1, pp.3–62. New York: Appleton-Century-Crofts.

———. 1969d. Helminth "Zoonoses": An analysis. *Helminthological Abstracts* **38**:333–51.

———. 1969e. Studies on ascaridoid nematodes in pythons: Speciation of *Ophidascaris* in the Orential and Australian regions. *Parasitology* **59**:937–59.

———. 1970a. Studies on ascaridoid nematodes in pythons: The life-history and development of *Ophidascaris moreliae* in Australian pythons. *Parasitology* **60**:97–122.

———. 1970b. Studies on ascaridoid nematodes in pythons: The life-history and development of *Polydelphis anoura* in Australian pythons. *Parasitology* **60**:375–97.

———. 1970c. *Baylisascaris tasmaniensis* sp.nov. in marsupial carnivores: Heirloom or souvenir? *Parasitology* **61**:75–86.

———. 1970d. Studies on ascaridoid nematodes in pythons: An outline. In *H.D. Srivastava commemoration volume*, ed. Kunwar Suresh Singh and B.K. Tandan, 417–28. Izatnagar, U.P. India: Division of Parasitology, Indian Veterinary Research Institute.

———. 1971a. Speciation and development in the genus *Lagochilascaris*. *Parasitology* **62**:71–112.

———. 1971b. A new genus and species of ascaridoid nematode from the marsupial wolf (*Thylacinus cynocephalus*). *Parasitology* **63**:37–43.

———. 1971c. A note on *Lagochilascaris* from the cat in Argentina. *Parasitology* **63**:45–48.

———. 1972a. *Cotylascaris thylacini* a synonym of *Ascaridia columbae*. *Parasitology* **64**:331–32.

———. 1972b. Obituary. Dr. M. Josephine Mackerras. *International Journal for Parasitology* **2**:181–85.

———. 1972c. Nematoda. In *A textbook of zoology* by T. Jeffery Parker and William A. Haswell. 7th ed. rev., ed. A. J. Marshall and W.D. Williams, vol. 1, pp.235–65. London: Macmillan.

———. 1972d. *Toxocara vajrasthirae* sp.nov. from the hog-badger (*Arctonyx collaris*) of Thailand. *Parasitology* **65**:491–98.

———. 1973a. The ascaridoid nematodes of rodents with a redescription of *Porrocaecum ratti*. *Parasitology* **66**:367–80.

Sprent, J.F.A., Lamina, J. and McKeown, Ann. 1973b. Observations on migratory behaviour and development of *Baylisascaris tasmaniensis*. *Parasitology* **67**:67–83.

Sprent, J.F.A. 1973c. Studies on ascaridoid nematodes of pythons: Two new species from New Guinea. *Parasitology* **67**:229–45.

———. 1974. Evolution of ascaridoid nematodes in amphibia and reptiles. *Proceedings, Third International Congress of Parasitology* München, 25–31 August 1974. **1**:475.

———. 1977a. Studies on ascaridoid nematodes in pythons: A résumé. In *Excerta Parasitológica en memoria del doctor Eduardo Caballero y Caballero* 4, pp.477–85. Instituto de Biologia, Mexico, Publicaciones Especiales.

———. 1977b. Ascaridoid nematodes of amphibians and reptiles: *Dujardinascaris*. *Journal of Helminthology* **51**:253–87.

———. 1977c. Ascaridoid nematodes of amphibians and reptiles: *Sulcascaris. Journal of Helminthology* **51**:379–87.

Sprent, J.F.A. and Jones, H.I. 1977d. *Toxocariasis. Australian Family Physician* **6**:1519–25.

Sprent, J.F.A. 1978a. Ascaridoid nematodes of amphibians and reptiles: *Goezia. Journal of Helminthology* **52**:91–98.

———. 1978b. Ascaridoid nematodes of amphibians and reptiles: *Paraheterotyphlum. Journal of Helminthology* **52**:163–70.

———. 1978c. Ascaridoid nematodes of amphibians and reptiles: *Gedoelstascaris* n.g. and *Ortleppascaris* n.g. *Journal of Helminthology* **52**:261–82.

———. 1978d. Ascaridoid nematodes of amphibians and reptiles: *Polydelphis, Travassosascaris* n.g. and *Hexametra. Journal of Helminthology* **52**:355–84.

———. 1979a. Ascaridoid nematodes of amphibians and reptiles: *Multicaecum* and *Brevimulticaecum. Journal of Helminthology* **53**:91–116.

———. 1979b. Ascaridoid nematodes of amphibians and reptiles: *Terranova. Journal of Helminthology* **53**:265–82.

Sprent, J.F.A. and McKeown, E.A. 1979c. Studies on ascaridoid nematodes in pythons: Development in the definitive host. In *Dynamic aspects of host-parasite relationships*, ed. A. Zuckerman, **3**:155–76. Jerusalem: Israel Universities Press.

Sprent, J.F.A. 1980a. Ascaridoid nematodes of amphibians and reptiles: *Angusticaecum* and *Krefftascaris* n.g. *Journal of Helminthology* **54**:55–73.

———. 1980b. Ascaridoid nematodes of sirenians — The Heterocheilinae redefined. *Journal of Helminthology* **54**:309–27.

———. 1982a. Host-parasite relationships of ascaridoid nematodes and their vertebrate hosts in time and space. In *Second Symposium on Host Specificity among Parasites of Vertebrates, 13–17 April, 1981. Memoires du Muséum National d'Histoire Naturelle Serie A, Zoologie, 1982* **123**:255–63.

———. 1982b. Ascaridoid nematodes of South American mammals, with a definition of a new genus. *Journal of Helminthology* **56**:275–95.

———. 1983a. Observations on the systematics of ascaridoid nematodes. In *Systematics Association Special Volume No. 22, Concepts in Nematode Systematics*, ed. A.R. Stone, H.M. Platt and L.F. Khalil, 303–19. London: Academic Press.

———. 1983b. Ascaridoid nematodes of sirenians — A new species in the Senegal manatee. *Journal of Helminthology* **57**:69–76.

———. 1983c. Ascaridoid nematodes of amphibians and reptiles: *Typhlophorus, Hartwichia* and *Trispiculascaris. Journal of Helminthology* **57**:179–89.

———. 1983d. Ascaridoid nematodes of amphibians and reptiles: *Freitasascaris* n.g. *Journal of Helminthology* **57**:283–90.
———. 1984. Ascaridoid nematodes. In *Diseases of amphibians and reptiles*, ed. G.L. Hoff, F.L. Frye and E.R. Jacobson, 219–45. New York: Plenum Press.
———. 1985a. Ascaridoid nematodes of amphibians and reptiles: *Orneoascaris*. *Annales de Parasitologie Humaine et Comparée* **60**:33–55.
———. 1985b. Ascaridoid nematodes of amphibians and reptiles: *Seuratascaris* n.g. *Annales de Parasitologie Humaine et Comparée* **60**:231–46.
———. 1985c. Ascaridoid nematodes of amphibians and reptiles: *Raillietascaris* n.g. *Annales de Parasitologie Humaine et Comparée* *60*:601–11.
———. 1985d. Ronald Harry Wharton 1923–83. *Historical Records of Australian Science* **6**:293–301.